Here's what others have said about Allan Nation's Pa$ture Profit$ With Stocker Cattle:

"If this book doesn't create controversy and spark some soul searching, particularly in the cattle business, both beef and dairy, it isn't for lack of effort on the part of the author."

Draft Horse Journal

"Nation writes that (Gordon) Hazard hasn't lost money on stocker cattle in 40 years at the business. How he's managed it makes good reading for anyone even slightly interested in harvesting grass with livestock."

Livestock Weekly

"Filled with good information on running a stocker cattle business, the book provides information that can be applied to improve business management techniques for other businesses as well."

Small Farm Today

"This book is everything they NEVER taught us in college a must for anyone in understanding the economics of agriculture."

Trinity Ranch, Fayette, Missouri

Pa$ture Profit$

With $tocker Cattle

by

Allan Nation

Revised Edition

A division of Mississippi Valley Publishing Corp.
Ridgeland, Mississippi

First printing July 1992
Second printing January 1993
Third printing April 1994
Fourth printing July 1996
Revised edition February 1998
Second printing May 2001
Third printing April 2003
Fourth printing October 2005
Fifth printing January 2009
Sixth printing July 2013

Copyright   1992, 1998 by Allan Nation

Library of Congress Cataloging-in-Publication
Nation, Allan.
   Pasture Profits with Stocker Cattle / by Allan Nation. -- Rev. ed.
      p. cm.
   On t.p. all "[s}" appear as the dollar sign.
   Includes index.
   ISBN: 0-9632460-7-0
   1. Cattle--Economic aspects. 2. Pastures--Economic aspects.
   I. Title.
   HD9433.A2N38   1998
   636.2' 13--DC21   97-41057
                                                              CIP

Cover Design by Steve Ericson, Ridgeland, MS

Manufactured in the United States of America.

This book is printed on recycled paper.

Table Of Contents

This book is dedicated to Gordon Hazard, the best teacher a stocker-grazier could ever have.

Foreword

"Any enterprise built by wise planning, becomes strong through common sense, and profits wonderfully by keeping abreast of the facts." Proverbs 24:3-4

T his book is dedicated to and primarily inspired by Gordon Hazard of West Point, Mississippi. Gordon was the first grazier I ever met who confidently stated that he made money with beef cattle year-in and year-out regardless of the price of cattle. This statement started me on a decade-long study of Gordon's Mississippi prairie stocker operation. You'll find a story about it in Chapter One. Most of the chapters in this book are a more in-depth examination of segments of his operation.

While Gordon has skillfully meshed his production methods with his particular climate and soil resources, I know other graziers who produce greater gains per head and per acre, grow more grass, and are more sophisticated in their grazing techniques. But they are nowhere near as successful as Gordon. I now believe it is Gordon's basic financial structure that is the real secret of his success, and that most grazing operations that fail do so because their underlying basic financial structures were unsound. Unfortunately, these "sure suicide" designs are the ones most often studied and reported on by the general farm press.

Therefore, the primary purpose of this book is to offer an alternative path to wealth. This book is written for those who want to get rich slowly, but surely, with a minimum of financial risk. Much of what I am going to say is going to fly in the face of everything you may have been taught about making money in the cattle business. Some of it will probably make you mad. Hopefully, all of it will make you think.

The basic financial principles of this book are applicable everywhere in every type of climate and situation, however, most of the specific production practices illustrated in this book are for humid climates.

Because the number of seed companies, grazier products, suppliers and grazing consultants continually changes, I have made no attempt to list those here. For the most up to date information consult the **Stockman Grass Farmer** magazine. A free sample copy can be ordered from the back of this book.

The Grass Garden

I grew up on a beef cattle ranch that utilized the levee and batture of the Mississippi River for grazing. My father had gone to college and earned a degree in horticulture. His farming goal was originally to have a very intensively-managed market-garden, but the opportunity to lease the grazing rights to the levee for the measly sum of one dollar per mile quickly led us into the beef cattle business. Each mile probably included 500 to 1000 acres of grazeable resource during normal river stages and could comfortably support 100 cows year-round with no hay.

We also purchased a small 40-acre dairy adjacent to the levee to give us a secure base of operation. We quickly phased out the dairy operation and put the whole farm into beef. This precipitous shift from dairy to all beef was probably an economic mistake, for my Dad brought his love of agronomic diddling with him from his horticulture days. His mental image of a pasture was a grass garden, and we were forever planting the very latest in grasses and legumes, even though our very low land costs allowed a far less intensive approach. Unfortunately, the payoff for this extra agronomic effort never seemed to quite make it to the bank. We quickly ran into the "pushing on a string" syndrome so common with beef cow-calf improvements. Over the years our operation evolved into the typical low cash and labor input extensive ranch with a corresponding low stocking rate and return per acre.

What we didn't realize was that if our goal was to grow a grass garden with lots of inputs per acre, we had to have a class of animal that could maximize these inputs; and the beef cow was the wrong class to do this.

The true beef breeds are bred to get fat rather than increase milk production like a dairy cow. This enables them to fatten during the summer and over-winter on low quality forages supplemented from their own body fat reserves. The beef cow has been described as a "bulk grass scavenger." She requires large amounts of grass, but this grass can be of relatively low quality once she has rebred. After a cow has raised her calf to four months of age, she serves primarily as a protein supplement to the calf. Her milk contributes less than a half pound a day to the calf's gain. The calf with its small mouth can select a much higher quality diet than its mother. With its low body mass, virtually all of this quality forage can be funneled into salable growth. The cow, however, with her large body mass to maintain, consumes copious amounts of grass just to supplement the calf at its side and grow the one in her womb. And the bigger the cow's body phenotype the less she can produce per acre. There are no free lunches in nature.

The beef cow was genetically selected to cushion nature's blows from her own body fat reserves, and the dairy cow was bred to give it all in the milk pail with the understanding that man would have to provide for her in the winter. We can produce calves, often very large ones, by crossing beef cows with dairy breeds. We must tremendously increase the forage input because we have removed the easy-fattening and self-supplementing effect of the beef genetics. The beef cow has one role to play in harvesting forage in North America and the dairy cow has another.

Several studies have shown that if we consider the entire ten-year span of the cattle price cycle, it is the low-input, extensive beef cow-calf operation that is the most economically sustainable. In gross economic terms, the worth of the beef cow is indirectly proportional to the value we put on the grass. If we start increasing the value (quality) of the grass, the economic value of the beef cow begins to fall, and the dairy cow rises.

Since a great many graziers are unwilling to commit to the physical and emotional constraints of seasonal grass dairy production, despite its exceptional returns per acre, the class of beef animals that most closely approximates the economic returns of the lactating dairy cow is the weaned beef animal or stocker calf. The weaned calf, due to its lower bodyweight and subsequent maintenance cost, can produce a much higher gain per acre than a cow. In fact, the lower the bodyweight of the calf the higher the potential gain per acre.

The restraining factor is that the grass for young stocker cattle must be of high quality. Higher quality grass in the higher rainfall climates can only be achieved with management inputs. When we start thinking stocker cattle we must start thinking in terms of dairy quality (and cost) forages. The beef stocker calf competes more with the dairy cow for forage resources than with its mother.

I believe a correctly structured beef cow-calf operation can be as profitable as a dairy or stocker operation on a return-to-capital basis, but not on a return-per-acre basis. However, this does not necessarily disadvantage the beef cow. North America has a huge resource of inexpensive beef-cow-quality forage in its rangeland, forests and crop residues. There is no long term economic pressure to agronomically increase beef cow output per acre when so many of these low cost forage resources are presently going unused.

When viewed from the standpoint of utilizing our total forage resources, the most profitable beef cattle structure combines an extensively grazed cow herd with an intensively grazed stocker operation. Ironically, the whole emphasis of animal science over the last 25 years has been to "cut-out" the stocker phase by increasing weaning weights and decreasing the weights the cattle would go on feed. This has resulted in increasing costs in both the cow-calf phase and the finishing phase and has, in effect, "cut-out" the most profitable phase of beef cattle production.

Hopefully, as more and more ranchers start to think of themselves as grass farmers whose sole job is to harvest the forage resources of this continent, they will be able to resist these misguided efforts. They will then come to realize that, in fact, forage harvest efficiency and profitability is increased by decreasing both the time spent on the teat and the time spent on feed.

Chapter 1

Gordon Hazard — The Guru Of Grass

"Wisdom is enshrined in the hearts of common men, but it must shout loudly before fools will hear it." Proverbs 14:33

Dawn. Early August. Its payday on the prairie. Carefully hidden in a valley area out of sight of the load-out corrals, five murmuring cattle trucks wait. They have come to haul away five potloads of steers that have hit Gordon Hazard's 800 lbs payweight target.

Guests are warned to stay out of sight and be quiet. The truck drivers know from past experience that if they make one untoward noise they will be sent on their way empty.

As the huge red sun first peaks over the tree line, from out of the morning mist there appears a tiny pickup truck with a huge black and white Brahman/Holstein steer slowly walking beside it. Behind this steer are 300 yearling steers bucking, snorting and prancing. A single rider on horseback slowly rides in a zig-zag pattern from one side of the herd to the other to keep the tardy from straggling too much. There is absolute silence other than the sounds the cattle themselves make.

The cattle enter a fenced wing that extends outward for nearly a quarter of a mile from the loadout corral. This wing narrows as it nears the fence line then turns in a sharp 45 degree angle to funnel to the corral gate. Turning the corner, Gordon speeds up, gets out of the truck, and walks into the open corral carrying a sack of range cubes. At the back Gordon pours out a small pile of range cubes for the big steer.

Seeing his treat, the big steer quickens his speed and gallops into the corral toward the pile of range cubes with all the steers that are to be loaded close on his tail. Gordon slips out a back gate and slowly walks around to the front gate, quietly closing it. He then whistles to the trucks and one by one they back up to the loading ramp. With the help of only one other person, Gordon loads a truck every 25 minutes.

After the steers are loaded, he walks over and scratches Ug, his pet lead steer, behind the ears and turns him out of the corral with a parting slap on his rump. The big steer shakes his head and trots back to his shady thicket at the back of the paddock, his day's work done.

Very impressive. No hassle. No noise. No stress. And most important of all, very little shrink. "I'll let a steer take one poop on my nickel but not two," Gordon grins. It is also very planned.

The steers loaded this day were sorted into uniform potloads by Gordon weeks earlier and sold to a feedlot buyer while still on the pasture. The pasture rotation was planned so that this group of steers would be in the loadout corral paddock on the day they were to be shipped.

All Any Grazier Needs

Gordon Hazard is famous throughout the Mississippi prairie country for both his innovativeness and his tightfisted ways. A few old wooden self-feeders and a 1952 model portable squeeze chute that would put most 20 cow part-timers to shame mark his receiving lots. He thinks horses, tractors and hay balers are for hiring not owning, and he will quickly tell you that.

Gordon lists his complete equipment inventory for his 1800 head stocker operation as one pickup, three hammers and a fence stretcher. "If times get tough, I sell one of the hammers," he laughs.

Such parsimony is not of necessity, but of priority. Gordon

Hazard lives in a huge Civil War-era mansion and readily admits to making around a quarter of a million dollars a year with his all-grass farm near West Point, Mississippi. "I'm not cheap. I'm tight," he explains. "I can squeeze a nickel so hard the Indian on one side rides the Buffalo on the other, but I'm always willing to spend a dollar to make two."

He describes himself as a child of the Depression. He is the son of an Iowa crop farmer who moved South and realized too late the financial trap plowing the shallow black soils of the Mississippi prairie always lead to. Seeing his father lose everything left Gordon with a strong skepticism about crops and plows. "I saw my Daddy go broke fighting grass on this prairie when grass is what the land wants to grow," he said.

Turning this grass into beef for 40 years has allowed Gordon to accumulate some 3000 acres that are free and clear along with a totally paid-for base herd of some 1800 head of stocker cattle. He is enthusiastic about grass farming and is always willing to share the secrets of his success with anyone with a genuine interest.

One of his biggest secrets is his 2800 lb Brahman/Holstein steer named Ug, which is short for ugly. Ug serves as the head cowboy for Gordon. The huge steer is of a pushy nature and quickly dominates the stocker cattle who are a fraction of his size. "Old Ug can put a group of steers in any pen or trailer with a minimum of fuss for a handful of range cubes," Gordon said. "One trained lead steer can replace four to five cowboys on horses."

One year as a test of Ug's worth as a shrink-saver, a comparison was made with a neighboring grazing operation in rounding up and loading four loads of cattle from each farm. The neighbor used four cowboys on horseback and Gordon just used Ug. The difference in shrink on the four loads when weighed in at the feedlot was a whopping $9000 in favor of Ug!

When Gordon makes his ranch rounds in his "little pickup" old Ug goes along with his right flank firmly pressed against the left side of the truck. Steering with one hand, Gordon scratches the big steer with his free hand through the window as he drives slowly along observing his cattle. He frequently lends out old Ug to his neighbors to get their cattle up and thinks all graziers are missing the boat by not getting a dominant lead steer.

Profit in Optimizing Resources

Gordon's basic attitude is "If it don't fit, don't force it." He said his program was built on optimizing rather than maximizing gain per head or per acre. "I'm happy with a pound to a pound and a third average daily gain over the course of a 365 day year."

He winters his cattle using fall stockpiled fescue and no hay and free-choice salt-limited cottonseed meal and selenium. He nitrates his pastures with 34 lbs of nitrogen (N) in the fall to accumulate growth and then rations this out over the winter. He currently has his 3000 acres subdivided into paddocks of approximately 80 acres in size.

"My program is based upon putting 300 to 400 pounds of gain on an animal as cheaply as possible. I winter my cattle cheap and try to maximize compensatory gain in the spring and summer. My cattle only gain a quarter to a third of a pound a day in January and December, over two pounds in the spring, and a pound and a half in the summer and fall."

He said his son, Mark, utilizes fungus-free fescue in his nearby operation and averages between 1.5 and 1.7 lbs a day per year. All new grass plantings are planted to fungus-free Kentucky 31 fescue. In Gordon's climate and heavy soils, warm-season dallisgrass naturally volunteers if rotational grazing is utilized.

Gordon said that with fenced pasture land selling for between $300 to $500 an acre (up from $150 an acre in the mid 1980s), he had put a pencil to it and found that buying land was cheaper than fertilizer. Dropping from a steer to the acre, he said he now utilized between an acre and an acre and a half to run a steer a full year on his all-grass, no hay system. He does keep an old tumble-down barn filled with hay in case of a weather emergency but said he hadn't fed any on the pastures for over 10 years.

He does use a lot of high quality Johnsongrass hay in his receiving lot. The hay is purchased "delivered in the pasture" to avoid the necessity for a tractor.

Due to the heavy clay soils and their tendency to pug badly in the winter, he set-stocks (no rotation) in the winter. The stocking rate is determined by the amount of fescue in the paddock.

He tries to have his stockpiled fescue grazed to a very low residual by early March to release the companion warm-season dallisgrass, white clover and lespedeza. By not nitrating in the spring, his fescue is soon well integrated by the dallis and clover. This dilution holds summer fescue toxicity problems to a minimum. In March, he puts his cattle together into herds of approximately 500 head. Each 500-head herd will have at least 10 paddocks to rotate through.

Gordon said he liked the stocker business better than cow-calf because it was much more flexible and fast-paced. Far from being a high-risk venture, he said stockering was virtually a no-risk enterprise if you could grow quality pasture, and if you structured it right financially. He said at the heart of every successful stocker operation there should be a production system capable of producing at least 300 lbs of cheap gain per head. A large amount of low-cost gain provides a cushion to protect you from any unforeseen market wrecks.

The second most important aspect of a successful stocker operation is to always try to buy your replacements as soon as possible after selling or booking your outgoing heavy yearlings. Because Gordon is in the Deep South his sale period is limited to the second and third quarters of the year because of the widespread feedlot prejudice against feeding Southern cattle in cold weather. Carefully avoiding the normally depressed May market, he plans to ship his steers in early August. However, he frequently sells them in February and March via a forward delivery contract for August delivery.

This contract price is based upon the feeder cattle futures for the delivery month minus the local basis. Basis primarily reflects the transportation cost between where your cattle are and the primary cattle feeding areas of the country. It also reflects demand for cattle from your particular region. With a guaranteed basis

Gordon can watch the national cattle futures and be able to figure his net price if he contracted at that instant. Most large order-buying firms and large commercial feedyards offer such basis and forward price contracts. Gordon uses Prairie Livestock Inc. in West Point, Mississippi.

The basis discount for Southern cattle is very wide during the cold months of the year but narrows during the hot months. This basis discount must be negotiated with whomever you are contracting, agreed to and a "guaranteed basis" contract signed. At the time of the contracting, Gordon receives a down payment of $40 a head. He normally sells his cattle in a staggered pattern of three to five loads per contract.

This contracted price then becomes the benchmark price against which he buys his replacements. While it may not always be possible to buy back immediately upon selling your cattle, the key point to remember is that you have no profits until your inventory has been replaced. As you will see later in this book, this is particularly critical during the rising price phase of the cattle cycle.

Flexible Stocking Rate

A New Zealand analysis found that the stocking rate used was one of the greatest variables in net profitability. Gordon deliberately plans to have his cattle small and few in number in the winter when grass is short and scarce, and big and numerous in the late spring and summer when the grass is plentiful and cheap.

By contracting in February and March (for August delivery) he feels safe buying their replacements early to "load up" his pastures to utilize the rapid spring and early summer grass growth. In effect, he still has all of the previous year's cattle (already sold via contract) and a portion of the next year's cattle on the pasture at the same time. "I don't mind summering cattle twice, but I only want to winter them once."

Gordon found it paid big dividends to stay flexible as to the type and breed of calves to buy. He said there was frequently more money to be made on discounted breeds than on Choice Number Ones. He found during periods of high prices he could sell any breed of cattle as long as they were uniform and in a truckload lot.

"I've given up trying to figure cattle feeders, " Gordon said. "I had a cattle feeder from Iowa here yesterday and he bought one load of straight English and one load of straight Brahmas. I've got other feeders who will only take Holsteins and others who only want shorts. There's a market somewhere for every type and breed, so I'd advise you to not get too hung up on the kind of cattle you buy. Also, don't overlook the heavily discounted fat calf in the fall. His fat shows he is a good doer and will winter well."

This genetic flexibility becomes much more limited during the low price period of the cattle cycle, and feedyard buyers can afford to be much more picky. As a result, Gordon buys higher quality cattle during the falling price period and lower quality cattle during the rising and high priced period. This decrease in quality is necessary to avoid excessive negative marketing margin (rollback).

He also maintains margin by buying cattle lighter and grazing them longer as the prices rise. This also helps to maintain profit margins per head. See an explanation of this phenomenon in the chapter "Cash Flow Versus Inventory Valuation."

While Gordon personally knows many of the cattle feeders who buy his cattle, he sells his cattle through Prairie Livestock in West Point to avoid credit problems. "If their check bounces they (Prairie Livestock) have to go get the cattle, not me."

A Low Stocker Death Loss Starts With a Sharp Eye

Several years ago I asked Gordon Hazard what the average death loss was on his annual purchase of 1800 stocker cattle. "One or two," he replied.

"One or two percent?" I asked.

Cocking his head to the side and giving me that wide-eyed "Were you born yesterday?" look that only Gordon can give, he answered again, "No, one or two calves!"

Therein lies one of the greatest lessons Gordon can teach about stockers: Death loss is not inherent to the stocker business, "southern" cattle or "salebarn calves." Death loss is controllable, and controlling it is a skill that can be learned and perfected.

Facility Design

Gordon said your receiving lot facility design was all important. He said turning a load of freshly weaned calves into a several-hundred-acre wheat field or pasture was a sure-fire recipe for disaster. You have to have the ability to observe your calves closely and then slowly walk the sick ones up for treatment. "Think of these calves as sick children," he said. "If your child was sick would you rope him and drag him to the hospital?"

Gordon's 30-day receiving program starts in a 5-acre grass receiving lot with tight woven-wire fences. He said tight, calf-proof fences are absolutely necessary with home-seeking, fresh-weaned calves. The water trough and self-feeder are set perpendicular to the fence so that calves will naturally bump into them as they walk the fence. The grass in this lot is kept short both to keep quality high and to prevent lying down sick calves from being overlooked. Free-choice hay is available in ring feeders at all times.

While Gordon's cattle never receive hay again after the completion of the receiving program, he said hay is critical to get rumen bacteria going again after stress and being off feed. This hay is purchased delivered in the ring feeders. By buying his hay in this way he does not have to own a tractor to put the large round bales in the ring feeders.

The receiving lot has a wing extending into the lot that gradually narrows in a semi-circular fashion to the corral and squeeze chute. This allows calves to be easily walked into the corral by one man on foot. The corral is divided into a holding pen and two sick pens. The working chute divides the holding pen and the two sick pens. This area of the corral is roofed to provide a dry and shady working environment. Worked calves can be turned into the sick pens or back into the receiving lot from the squeeze chute. Gordon said it is important that the funnel alley leading to the squeeze chute not be over 20-inches wide. This will prevent small calves from being able to turn around.

Encircling the receiving lot on three sides are four 30-acre grass "traps." These traps also have wooden self-feeders and roundbale hay feeding rings. All these traps have angled fence wings and walk-back lanes to the receiving lot to make it easy to move calves from these traps to the receiving facility for treatment.

In-coming calves move through these "traps" one week at a time. The first two traps have unlimited free-choice feed, but salt is added in the next two to encourage the calves to self-wean to the pasture. The first three traps' feed has a coccidiostat added. The last trap doesn't.

All feed manufactured for Gordon's farm has selenium added at the maximum legal level. He has found that selenium works to both offset stress and to minimize the effect of fescue toxins. He said he has had only two calves come down with fescue foot since he started supplementing with selenium several years ago. All feed is purchased from a local feed mill delivered in the trap self-feeders. These self-feeders are always located near the fence and a hard-surface road, so that they can be filled without the feed truck driver having to bring the truck into the trap. The feed is augured into the self-feeders from over the fence.

The Procedure

Gordon's receiving program lasts for 30 days, the first 17 days of which he said are the most critical. He said to keep in mind that the incubation period on most calf diseases is 7 to 10 days. This means you should not expect any of your calves to arrive sick. If they do, they are "stale" calves and have been hanging around the order-buyer's place too long. He recommends that you reject the calves and send them back to the order-buyer.

Gordon said it is necessary to start with calves relatively fresh from their home pastures to keep your death losses at the absolute minimum. He said calves needed to be vaccinated for whatever are the major health problems in your area before they arrive at your place. He said a good large-animal vet in your area could tell you what these are and a good relationship with a vet was invaluable in keeping death losses low.

"Be sure and get a vet who has had experience with stocker cattle," he said. "The diseases of cattle are entirely different from those of dogs and cats."

Gordon said the smaller the group of incoming calves you can deal with at one time, the better. He prefers to not handle over one truckload lot of calves at a time. He said he wants his calves delivered at his farm between 6 and 7 AM. Do not allow calves to be delivered at night when you cannot clearly see what the buyer has purchased and the health of the calves. If you are going to send calves back, you want to do it on the same truck that brought them. He also said to not allow weekday purchased calves to stay at the order-buyer over the weekend as the calves' virus clock was running. "Virus bacteria have a 7 day incubation period. It is critical that we not lose that window of opportunity," he said.

Observation Is the Key

Gordon castrates the calves immediately upon arrival. However, he does not dehorn until after the calves have completed their 30 day receiving program. He believes dehorning is just too much stress for an incoming calf and it is not necessary to get in a hurry about it with lightweight calves. "They are going to be here a year. We'll dehorn them later when we get them up to worm them," he said. He applies what he calls the "Oklahoma Bop" that leaves approximately an inch of horn remaining and does not expose the sinuses to the air.

He said after your calves arrive to just sit for 15 minutes and watch the calves from a pickup truck parked inside the receiving lot. He said to watch for calves that aren't feeling well. He said you can tell a calf that is going to get sick by the way he carries himself. If his eyes aren't bright and his head isn't higher than his shoulders you've got a problem coming and the sooner you can get on it the better.

"I want to treat a calf the moment he thinks about getting sick. If he is sick when I treat him, I haven't done my job, " he said.

Conversely, he said there is no antibiotic that can make a well calf "weller" and Gordon doesn't believe in non-discriminatory mass treatment of calves with antibiotics. He also doesn't believe in thermometers as a judge of a calf's health either. He said some calves naturally have a higher than normal body temperature.

"You've got to learn to trust your eye," he said.

Separating calves who don't feel good from those who do is critical, because the calves who feel good will beat up on those who don't. Gordon said a sick calf can't compete with well calves for feed and water and this can start a downward death spiral. It is equally important that treated calves not be turned back in with the healthy calves for the above reason and because the sick calf could make the others sick. "A calf that gets sick on my farm will be kept separate from the others for 30 days," he said. "The biggest health wrecks we've seen in our stocker school alumni have come from graziers who just treated the sick calves and then turned them back in with the rest of the calves."

Gordon lives in town and typically arrives at his receiving lot before dawn. He then sits in his truck for two hours closely observing the newly treated cattle. "I want to be there when the calves first stand up in the morning. I want to see a calf stand up, stretch and shake himself awake. You can best see the health of a calf at dawn."

He said after five days, any calf that is gaunt and doesn't appear to have been eating well will be pulled and treated with an antibiotic.

What to Look For

☆ Hunchbacks on older cows indicates low calcium or low selenium.

☆ Animals low in selenium will scour. Strung-out manure indicates low selenium.

☆ Dung on the tail indicates worms. Normally animals will not dung on the tail.

☆ Manure on each side of the butt indicates coccidiosis. An itchy butt is also a sign of this.

☆ Healthy animals hold their heads up, have bright, open eyes, and a nice sheen to their hair coat.

☆ If any of your stock is bawling, something is wrong. They need water or minerals. Check it out. Don't ignore their plea.

☆ Lack of minerals can make animals wild and crazy.

☆ Animals licking each other indicates lack of sodium.

☆ Animals developing an unusually large head indicates the need for cobalt.

☆ A brown tinge to black hair indicates a lack of copper. Hair should always have a sheen to it.

The First 17 Days Are Most Critical

Gordon said the first 17 days are the most critical with stocker calves. "A calf that hasn't gotten sick after 17 days probably isn't going to get sick," he said, "and you can start to relax a little. After 30 days, we have almost zero health problems and the new calves are turned in with one of the herds."

Gordon has his cattle sorted into roughly 500-head herds of similar weights. Each herd has its own set of 80-acre paddocks it rotates through. The cattle nearest to being sold are always given the best grass. In the spring the cattle are bunched into large herds and moved rapidly to keep the fescue short. This fast rotation keeps the fescue at high quality and allows the companion warm-season dallisgrass full sun. In the winter due to the extreme bogginess of the heavy clay prairie soils, the cattle are set-stocked on stockpiled fescue and are not rotated. No hay is ever fed after the cattle leave the receiving lot.

Gordon plants all new pastures with endophyte-free fescue. These pastures produce a half pound a day higher average daily gain than the infected pastures.

Worm Every 90 Days

In hot, humid climates Gordon said it is critical that stocker cattle are wormed every 90 days to keep gains high. This frequent worming also apparently helps offset the effects of fescue endophyte. He said he liked to use Ivomec Pour-on in the summer months as he found he received excellent fly control as a bonus.

Free choice, salt-limited (2 to 1) cottonseed meal and selenium is available to the cattle at all times as a selenium carrier. He said this supplemental feed costs him 5 cents a day per steer but has eliminated most of the health problems connected with fescue toxicity and has increased his average daily gain by over a quarter pound a day. Hazard said his total variable costs (excluding only land and interest on operating capital) produces gain at around 25 cents a pound year around. Gordon said his profit per head consistently ranges from $125 to $160 a head and his total return on investment is around 25 to 30 percent including land ownership. This high return is due to his having virtually no investment in depreciable assets.

He said the "unreasonable advantage" he had in Mississippi was that he could winter a steer for only $15. He said Southern graziers had to build their operations to maximize their winter advantages and that Northern graziers had to maximize their summer advantage. Gordon feels that the best way to do this is for the North and South to work together.

For example, May is traditionally the lowest priced month of the year for feeder cattle and August one of the highest. Gordon occasionally buys local cattle and has them contract grazed in the Kansas Flint Hills to take advantage of the forward margin frequently available on May purchased/August booked feeders. However, Gordon doesn't believe you can build a business on such short-term "deals."

Gordon said the secret to a high profit stocker business is to maximize cheap gain per head. He feels a grazier's gain goal must be at least 350 to 400 lbs per steer. "If you study the cost structure of stocker cattle, 85 percent of the total cash costs occur in the first 30 days you own the animal. Therefore, the longer you own the animal and the more weight you put on per head, the lower your overall cost of gain will be and the higher your profit."

Gordon will frequently book feeder cattle for summer delivery in February, and buy their replacements at that time to "load up" his pastures for their rapid spring growth phase. "A booked steer is as sold as one that's gone through the sale barn. There's no risk in buying a replacement against that price."

He buys calves year-round, depending upon their price and his grass conditions, however most are bought between late August and Christmas. He has never used the futures market and sees no use for it in his type of operation. "If the feeder cattle market falls out of bed, the stocker calf market will too. By buying your re-placements on the same market you sell on, you minimize the effect of marketing margin on your cash flow."

Pasture Profits

Gordon Hazard's basic operating philosophy is, "If it don't fit, don't force it." In the 20 years I have known Gordon I have been keeping a list of what I consider to be some basic universal truths illustrated in his operation that can be put to use anywhere in the world. We'll be examining these and other points in more detail throughout the book.

* He measures his economic well-being in terms of cash flow, not inventory valuation.

* He has an optimistic, pro-active attitude. He can see the profit that lies within what others consider adversity.

* He keeps his herd flexible in numbers, class and grades.

* He avoids buying anything that rusts, rots or depreciates, and invests only in livestock and land.

* He concentrates on growing, appreciating livestock. A fully grown, fully priced animal can do nothing but lose value.

* He has minimal machinery ownership — one small pickup. All hay is purchased. All machinery work is done by custom operators.

* His cattle are self-financed. Initial leverage is restricted to land. (Now paid for). All expansion is self-financed from retained profits. Charge your cattle for grazing your land even when it is paid for. Put this money in a future expansion fund.

* All land acquisitions should be capable of grazing an out-going truck-load lot of cattle (50,000 lbs).

* Using sell-buy marketing provides market risk management. Decisions based upon actual prices in real time. Only sell to repu-table, credit-worthy cattle buyers.

* His operation fits his local climate, grass and market.

* He keeps a low cost of gain.

* He maximizes use of compensatory gain. Cattle are sold after best gains.

* He has a large gain (350 to 450 lbs) per head.

* He keeps a low labor input per head — 20 hours a week for 1800 head.

* Forages are chosen to match the existing soil fertility.

* The operation is designed to produce minimal animal stress and shrink (use of a lead steer, etc.).

* The farm is located in a marginal rowcrop area rather than a ranching area. Cheap land is always available from bankrupted crop farmers.

* He knew when to quit growing. Bigger is not always better.

* Fast growing enterprises find it difficult to show a positive net cash flow. As Gordon's son, Mark Hazard said, "It is better to creep forward through accumulation, than leap forward through debt."

Chapter 2

People Make Calves Sick

"A good man is concerned for the welfare of his animals."
Proverbs 12:10

G ordon Hazard's operation is an example of how to keep death loss and stress low. How you treat your animals from the moment they arrive at your place until the minute they leave has a profound effect on both your financial and your animals' health.

I recommend that you sit down with a vet and work out a treatment program for sick stocker calves. However, I do not recommend that you rely strictly on a needle for disease control. With stocker calves, an ounce of prevention is worth a pound of cure out of a medicine bottle.

Most disease in stocker calves is caused by stress. Stocker calves are subjected to many stresses that are beyond our control. But as stocker calf receivers, we must concentrate on relieving the stress put on the calf by the marketing system as quickly as possible.

Animal behaviorist, Bud Williams, told me the first requirement for a low death loss on stocker calves was to love your animals and your job. Williams works as a consultant at a feedyard in Alberta when he is not on the road teaching his "no stress" method of animal handling. He said showing up for work in the receiving yards in a bad mood was grounds for being sent home at his feedyard. "Our attitude is the most important thing in keeping stress off animals. If you are personally feeling stressed and uptight, you are going to produce sick calves," he said.

Williams believes that animals can sense our emotions and will mirror them. "A healthy calf is a happy calf. Animal health problems are created by somebody," he said. "Animals don't just get sick. They decide to get sick. If you don't want to be there with that calf he knows it and it bothers him."

Bud believes that almost all animal sickness is due to stress. He pointed out that weaning is not inherently stressful to a calf. "You don't see calves walking fences and bawling in nature and yet I have had dozens of ranchers tell me that fence-walking and bawling is a natural part of the weaning process."

Fence-weaning, whereby calves are gate-cut away from their mothers but can still see each other through an electric fence results in a stress-free weaning with no bawling and no sickness. Unfortunately for graziers who buy salebarn-weaned calves, such a gentle process has not been utilized. The calves arrive at the buyer's ranch in a highly stressed and agitated state of mind.

"The first thing you have to do is to get that stress off of them to keep them healthy," he said. This includes stopping the fence-walking. What you need to understand is that the calves don't want to be walking the fence. They are forced into it by their buddies. They want you to stop it just like kids want the teacher to break up the fights they get themselves into," he said.

Williams said that animal movement creates animal movement and animals will always move in the same direction unless prevented from doing so. In fence-walking the calf in front pulls along the calf behind who in turn pushes the calf in front until the whole group is totally exhausted. He said that walking alongside the calves in the same direction they were going would bring them to a halt. Taking the time to slowly walk with your calves and get them to focus their attention on you will distract them, which helps stop fence walking.

He said animals always want to know where you are. If you are alongside, they have to slow down to keep you in sight. If you then stop, they will stop, turn and look at you.

Conversely, walking against the direction the animals are moving will speed them up. Animals always want to go in the direction they are headed and they will escape the pressure you are putting on them by speeding up to get by you.

Never Get Behind an Animal

Bud said the absolutely highest stress you can put on ruminant animals is to get behind them where they can't see you. Ruminants have their eyes on the sides of their heads and can see almost 360 degrees around them. The one exception is a small area directly behind them. He said you should never be in this blind spot for more than an instant to keep animals calm.

"Cattle know they are a prey animal and they absolutely hate to be chased from behind. Humans, on the other hand, are a predator species and absolutely love to chase animals. Therein, lies the source of most of our animal health problems," he said.

Animals should always be driven from the side. He said attempting to drive an animal from behind will always turn the animal around so he can see you. This, in effect, produces the exact opposite result you are trying to produce as the animal will always go in the direction he is headed. Therefore an animal should always be facing the direction you want him to go before putting any pressure on him to get him to move.

If you want an animal to go through a gate, you should stand beside the gate. Once the animal is looking at you and facing the gate, you can step toward him and he will run by you through the open gate to escape your pressure.

When working the drag position behind a herd, the motion should always be back and forth rather than static so the animals always know where you are and don't turn around to try to see you.

He said the sole standard of productivity at most big feedyards and stocker operations seemed to be how fast one can work an animal. Bud said fast movement and noise puts stress on animals, and most vaccines will not work on stressed animals. He said animals should be worked slowly and in as close to a noise-free environment as possible. Receiving crews should be trained to avoid talking and to use hand signals. All squeeze chutes and hydraulic motors should be snubbed and muffled to work silently.

Drug Bill Has Fallen by $20,000 a Month

Williams said since he had put the stress reduction program in place at the feedyard in Canada, the death loss on newly arrived weaned calves had fallen to less than one percent. However, even more important, the feedyard's drug bill had fallen from $20,000 a month to around $200. "A low death loss may not increase your profits if you spend twice as much on drugs as the original death loss was costing you," he said.

Bud had noticed that as the drugs have gotten better the animal health problems have gotten worse because there is now an almost total reliance on the drugs to keep the animals healthy. "I am not a fan of doctoring. I would rather try to prevent sickness. The two best drugs for calves are feed and water. Most sick calves basically starve themselves to death. You've got to watch and make sure you have seen every calf eating."

He said it is often necessary to take the calves up to the feedbunk and show them where the feed is. Any calf that tries to run away should slowly be walked back to the feeder. Stay with the group until you have seen each of them feed.

Once the calves have eaten, slowly walk them to the waterer and stand with them until you are sure each has had a drink. Don't let any bully beat up or push around the smaller calves.

Stress relief extends to the pasture. When making paddock shifts with new calves, stay with them until they have settled down and have started grazing.

"It is more important that you pull and treat animals early than what drug you use. Close observation is the key to a low death loss. If you look at a calf twice, treat him. That second glance is your subconscious signal that something is wrong. The real trouble starts if we treat too late. We really want to doctor an animal the day before he is going to get sick. You need to develop the eye that allows you to see the calf that's going to be sick tomorrow.

"Always ask yourself why an animal got sick. Our job is not just to check for sick animals but to prevent the next one from getting sick."

Just as it is important to stop calves from fence-walking themselves into exhaustion, it is equally important to make sure they get exercise.

"I don't like to see calves lying around all day on a straw bed. Walking and exercise are good for them."

In the winter he takes the calves out of the feedyard every day and walks them over the frozen pastures for exercise. This exercise program greatly increased the feedyard's average daily gains and the animals' health.

Williams has little patience with stocker graziers who don't have the "time" to practice a good preventive animal health program and look closely at their calves every day. "If you don't have the time to take care of your animals, go do something else. You can't be too busy if you are going to own cattle."

An avowed animal lover, Bud said he actually preferred the company of animals to people. "Animals learn very fast. People don't. Most ranchers don't want you to teach them something new. They want you to show them how they can keep doing what they are doing and have a different result. This (hope for a different result without any change in action) has been given as the definition of insanity."

Pasture Profits

Williams said it is easy to work animals stress-free if you understood these concepts:
* The quicker you can relieve emotional stress on your animals, the better your chance of preventing ill health.
* Walking with the animals slows them down.
* Walking against the direction of the animal speeds them up.
* Animals like to be able to see you.

* Getting behind them in their blind spot sends them into total hysterics.
* Animals want to go in the direction they are headed.
* Animals like to move with other animals.
* Animals have very little patience, so you must have more.
* Slower is always faster with animals.
* Noise can stress animals just as much as a bad attitude from you.
* Animal stressors include:
>Pain from processing,
>Bullying by strange cattle,
>Wind,
>Wet,
>Cold,
>Heat,
>Dust,
>Mud,
>Noise.

Chapter 3

A Structure for Profit

"Develop your business first before building your house."
Proverbs 24:27

I t has been said that ranchers are land rich and cash poor. What most of us have failed to see is that the former creates the latter. I get countless requests from readers asking "Will X animal enterprise pay for the land?" That's the wrong goal for a ranch. The purpose of a ranch is not to buy and hold land but to make an operating profit. For a startup enterprise, land ownership can be like trying to swim a river with a wagon wheel tied around your neck. You may not drown, but you are guaranteed one heck of a struggle.

Land is an investment. Investments can only be made from after-tax operating profits. This requires the existence of a profitable tax-paying operation before one buys any land. However, the majority of new farmers and ranchers first buy land and then try to figure out an enterprise that will pay for it. There are people with significant off-farm income and careers who prefer to take on the role of land investor rather than livestock owner. These are the people a start-up grazing operation must seek out. The best place to start is your local bank's Trust Department.

In the future, ranching will closely resemble the urban office split of landlords and tenants and for the same reasons. Landlords have surplus capital. Tenants don't.

Renting Offers Cash Flow Advantages

Land principle payments are not deductible under USA tax law as a production expense. This puts a land-buying operation at a considerable cash-flow disadvantage to a land-leasing operation because lease and rent payments are 100% deductible as a production expense. Land rental payments tend to rise and fall with the profitability of farming and ranching, whereas, land prices tend to reflect the health of the off-farm economy.

Due to these considerable advantages to renting and leasing rather than owning land, high income producing farming and ranching operations try to keep the number of acres owned, versus those leased, relatively small. A good rule of thumb is to never have a mortgage more than twice the farm's annual taxable income.

Land is just like a large home. It is not that you shouldn't aspire toward owning one, but it should only be bought after one is wealthy. Not in anticipation of becoming wealthy.

If you separate ranching and land ownership into two different enterprises, what you pay for your farm or ranch is largely irrelevant. As in most real estate deals, the profitability of the land enterprise will be dependent upon subsequent price appreciation more than on-going operating income (rent).

A 1996 survey of land values in Texas found that farm and ranch land outside of urbanizing areas and undesirable for corporate hunting retreats have shown no increase in value since 1984. What many people are interpolating as an overall increase in farm and ranch land in their areas are the prices being paid for ten acre estates, not commercial-scale farms and ranches.

Large acreages, particularly those without highway access, can be very difficult to sell. Gordon got financially bailed out on his first farm by selling 4 acres of it for a truck stop. He later bought several farms and turned around and sold off the highway frontage to pay for them. He calls this the "doughnut farm" strategy.

"Your cows don't need a view of the road," he said.

Now, please realize that being able to see and take advantage of such situations require knowledge of the local real estate situation. This is why I say the land business must be seen and managed as a separate enterprise. Successful land speculation has nothing to do with farming and ranching. However, what Gordon did next does.

Rather than letting his cattle graze on these farms rent free, he charged each steer an annual "land charge" that he put in a special land expansion account. If a farm came up for sale and the account had enough money in it, Gordon, or rather as he describes it, "the cows" would buy it. If the account was short, "the cows" would pass. He financially structured his ranch to first create income, and then, to buy land from after-tax surplus cash flow.

All the animal enterprise should be expected to pay is what comparable farm or ranch land in your area rents for. In most agricultural areas this will be between 3 and 5 percent of the land's appraised value.

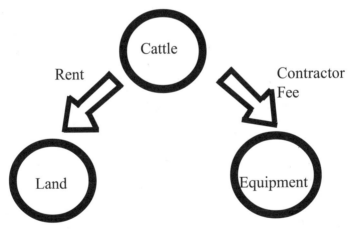

Machinery as a Separate Enterprise

The more we can allocate our costs into enterprise centers the more we can figure out where our profit problems may be coming from. What is most misunderstood about machinery ownership is that it is more costly the less it is used. We do not "save money" by using a machine less. European agricultural economists figure that a tractor must do at least 1000 hours of useful work a year to be cost-effective.

A good way to judge how much you really need a piece of machinery is to set your machinery up as a separate cost center that "sells" its work to the ranch at the going local contractor rate. (These can be obtained from your state's extension service.) However, keep in mind every "cost" is also a "benefit" that can possibly be sold to others. Selling your unused machinery time would not only lower the cost of your machinery ownership but can actually become a single profit center for your ranch. The point here is that all machines must work to be profitable. If a machine can't sell enough hours to pay its depreciation, its direct operating costs, and a competitive return to the capital tied up in it, you are better off buying that machinery work from others.

A lot of graziers tell me that their machines are already paid for and therefore don't have a cost. Yes, they do. This cost is the opportunity cost of the capital tied up in them. Even a junk $600 tractor is costing you big money if that same $600 could be invested in two to three steers that could earn you a 20% annual return. This is what Gordon Hazard means when he says he can't "afford" to own a tractor.

Seven Secrets To Successful Bootstrapping

1. Have a long term goal and outlook. Be willing to commit to 10 to 12 years of learning your trade and reinvesting your profits. You must always give something up to gain something new. A clean shave costs you a beard. You can't have it both ways!

2. Find a mentor. Most successful people are willing to tell you all they have learned in return for a little deference and respect. Avoid socializing with negative people. Avoid ranchers who are ranching primarily for fun and lifestyle. Remember, peer pressure is designed to dampen extraordinary efforts and results that would make the norm look bad. The best startup for a new young, bootstrapping grazier is a joint venture with an older successful, grazier.

3. Maintain a frugal personal lifestyle. Avoid expensive housing and machinery of all kinds. Direct as much of your capital

into livestock ownership as possible.

4. Have one team member with an off-farm income. Live on the off-farm income and reinvest all the dollars generated by the grazing enterprise into additional livestock. Eventually the off-farm job member can quit the off-farm job if so desired. However, keep in mind there is some financial stability in not having all of a family's income from one source.

5. Keep the size of your operation consistent with the level of your management skills.

6. Seek to substitute labor for capital whenever possible. As a boot-strapping grazier you are primarily selling your time via livestock management. Investments in "labor-saving" machinery are only cost-effective if that time saved is used to do more valuable work. Remember time is not money. Only money is money.

7. Concentrate your capital in animal ownership. Never seek more land than you can comfortably stock from your own capital resources. If you have inherited land, sell some and free up capital for investment in animals. Only after a paid-for herd of animals is capable of sustaining your family lifestyle and producing a capital surplus should land ownership even be considered.

Pasture Profits

* Wealth is created by a high return on capital.
* A high return in ranching is created by having the majority of your total investment in growing (or appreciating) livestock.
* Land ownership is a primary cause of low capital return to North American ranching.
* Machinery is only cost effective when it is doing enough useful work to offset its depreciation and produce a return on the capital invested in it.
* It is much easier to achieve an above average return on investment with cheap livestock than expensive livestock.

Chapter 4

Money Talks ...
Capital (Investment) Ideas

"Steady plodding brings prosperity; hasty speculation brings poverty." Proverbs 21:5

Mark Hazard is Gordon's son, a stocker grazier, and the president of First National Bank in West Point. His bank has loaned millions on stocker cattle and has never lost a dime on a grass-based farming enterprise. Mark largely attributes this to the way he requires graziers to structure their loans, and operations.

As a banker, Mark expects his customers to borrow less on their animals each year. "You should work on taking the bank out of the livestock portion of your operation as quickly as possible," he said. "Our policy is to require the beginning grazier to put up 25 percent of the purchase price of the cattle. We'll finance the rest initially, but we expect to put up only 60 percent the next year, and so on. We expect the profits to be reinvested in additional equity in the cattle until they are completely paid off."

Most beginning graziers Mark works with have off-farm jobs and are able to reinvest all of their cattle profits in livestock.

The key point, he emphasized, is that the initial investment should always be directed toward animals not land. "If you own your own cattle there's no way the market can take you out."

Young farmers often have to start out contract grazing for others to build equity and expertise, but the primary profit is in cattle ownership. Most cattle investors would prefer to put their cattle with someone who has his own cattle there as well. "It's too easy to stay in bed on a cold, sleety night if it's someone else's cattle out there." Mark said non-leveraged stocker cattle could pretty easily show a better than 30 percent annual return to the capital invested in them year in and year out, and were one of the best ways to build capital.

"If a grazier has paid for land, or land with a lot of equity in it, I usually encourage the grazier to re-leverage the land and pay off his livestock loan. This is good for the grazier in that it gives him a long term loan at a fixed rate and you don't have to consult your banker on every set of animals you buy. Bankers feel far more comfortable with long-term real estate loans than cattle loans."

Mark encourages his stocker clients to use sell-buy accounting whereby profits are not figured until a replacement animal is purchased. This form of accounting is much safer because the negative margin that must be absorbed is known at the time of purchase. "It may be that you have to buy back heifers or a lesser grade of cattle to keep your negative margins in line. A $10 to $15 negative margin is relatively easy to overcome but a minus $40 takes all the fun out of the stocker business," Mark said.

He started out grazing heifers exclusively. Heifers gain just as well as steers until both sexes are a year old. Then the steers run off and leave the heifers. As a result of slower heifer yearling gains, heifer graziers may want to sell their animals younger and lighter than steer graziers. He usually sells his heifers at 600 to 650 lbs.

Mark prefers lending money to graziers using management-intensive grazing because it indicates professionalism and guarantees that the cattle are being looked at on a daily basis. "The first thing I want to know is can you handle sick cattle? The second thing I want to know is do you know how to handle your grass?"

Patience, The Key To Wealth

Mark Hazard said the primary attitude needed for success in today's capital short world is patience, and gave the following examples:

☆ Patience to grow into the grazing business without major debt.

☆ Patience to keep from buying cattle that will have a high negative margin. The money always runs out before the cattle do.

☆ Patience not to follow someone else's hot new idea until it has been pasture proven.

☆ Patience to buy cattle lighter and graze them longer.

☆ Patience to go to heifers that can go as feeders or stay an extra year and become springers.

Cutting Costs In Grazing

The following examples are ways Mark and his father have found to cut costs in the grazing business:

☆ Own the cattle and cut 10 to 11 percent interest costs right off the top. If you own your cattle there is always a profit.

☆ If you are grazing in the cool weather months, English cattle will do better and are often cheaper to buy. Brahman crosses shiver in the winter but turn it on in the summer. Brahman crosses should be sold by August 15th. This will allow the animal to be dead or have a significant fat cover before winter.

☆ Match your fertilization to the amount of grass your cattle need. Don't mow grass that you can utilize.

☆ Avoid selling hay off your land. Put as much of your grass as possible through a gaining animal. Buy hay from your neighbors.

☆ Buy more cattle to avoid clipping spring pastures.

☆ Never spray pastures for aesthetics. Only control weeds where they are seriously affecting pasture quality and quantity.

☆ Avoid rust. A light truck and a stock trailer is all any grazier should need. Hire everything else done.

☆ A good intelligent lead steer can replace four cowboys and horses.

✮ Keep your cattle gentle by driving through them, putting out a few range cubes, leading them up the lane into the corral and feed a few cubes in the pen even when you are not going to do anything to them. Make friends with your animals and make them glad to see you.

✮ Know your cattle buyer. Is he honest and can he pay when the cattle are delivered? If the cattle are sold for future delivery, get enough money down to take care of a price drop. Having a load of steers 1000 miles away when the check bounces can cause major financial and heart problems.

✮ Avoid shrink. Cattle can lose two weeks of gain in two hours if they are treated rough and they get nervous. Standing overnight in a corral costs 8 percent in shrink. Four hours in a sale barn costs 4 percent.

Shrink

Time off feed and water	Liveweight loss %
1 hour	1.5
2 hours	2.5
4 hours	4
8 hours	6
12 hours	7

In all his years of grazing, Mark has only lost money on one set of cattle. That was when he decided to "upgrade" the quality of the cattle he was buying and put together the prettiest set of cattle he had ever seen. He quickly learned that pretty is not necessarily profitable. Now he follows his Dad's advice of only buying calves that would be going to "a better home than they have come from."

Pasture Profits

* Get your banker out of the livestock portion of your operation as quickly as possible.
* Leverage land not animals.
* Grow from retained profits rather than debt.
* Pasturing cattle for and with others saves capital.

Chapter 5

Cash Flow Versus Inventory Valuation

"Without knowledge even zeal is not good; and he who acts hastily blunders." Proverbs 19:2

Gordon's sell-buy system is known as replacement inventory accounting and is a cash flow accounting system which ignores paper losses from inventory valuation changes.

You are always working with two economic factors in ranching. You have an inventory value and you have cash flow. It might be helpful to think of these general business terms as your inventory being the worth of publicly traded stock, and cash flow as your business' operating profit.

The price of stock is largely controlled by the vagaries of the world economy and human emotions and is largely outside the purview of management. While it is important in terms of liquidation value and bragging rights, businesses don't go broke because of a drop in the price of their stock. They go broke because of the lack of cash flow. This is also true of stocker operations.

In October of 1987, the New York stock market fell 500 points in one day. The "losses" were estimated by the financial press as being in the hundreds of billions of dollars. In January of 1988, the New York stock market passed the high of October

1987. So, what became of the losses in the billions then? It is obvious that they were only experienced by those who chose that day to quit the stock market forever. Those who minded their knitting and held onto their investments experienced no loss at all.

In **Managing in Turbulent Times**, Peter Drucker repeatedly points out that paper "profits" and paper "losses" are largely illusory if viewed as a continuum. Like our stock market example above, they only become real if you choose to leave the game. If you choose to stay with the game, today's "profits" must be reinvested and the game played again tomorrow.

This is not as easy as it sounds. At the bottom of the market there will be a lot of wailing and gnashing of teeth about how no one is eating beef and other ghost stories and you must steel yourself to not fall victim to this negativism. This is why you need a well thought out strategy in writing. If you don't write it down you will be constantly changing it or making it up as you go along. This makes you extremely susceptible to negative reporting. The stocker market is just like riding a tiger. There is very little risk as long as you stay on his back. The more times you get off and try to get back on, the more likely it is that you will be eaten.

Inventory value is the net value of the animals when sold. 500 steers at 80 cents a pound have a greater liquidation value than 500 steers at 60 cents a pound. Whether steers are worth 80 cents or 60 cents is beyond the control of the grazier in a commodity-priced market. Like the value of a share of stock, the price of the stock is only really important when it is liquidated and the money used elsewhere. The fluctuations while we hold the stock are really not important.

The sell-buy accounting used by Gordon might be more correctly expressed as BUY, sell-buy, sell-buy, sell-buy, SELL. Like the share of stock, your inventory profit can only be determined on your last sale of the last set of cattle you graze before retirement. Until that point in time, what really matters is cash flow.

Gross cash flow is only created by the sale of an animal. Net cash flow is determined by the relationship between the animal we sell and the animal we buy to replace it.

It is very important that you understand the difference between inventory profit and loss and net cash flow increase or decrease. Here's why.

It is possible to have a rising inventory value but a negative cash flow.

It is also possible to have a falling inventory value whereby the books show you are losing money but you are actually experiencing an increased net cash flow.

Let me explain.

The second scenario occurs when your replacements' costs are falling faster than the price of your sale animals. This phenomenon happens during the falling price phase of the cattle cycle. However, on the up-cycle the reverse situation exists.

During a period of rising prices you can be showing huge inventory profits on the books due to rising cattle prices but be running out of cash because your replacement costs are going up in price even faster than your sale animals.

This dichotomy between falling cash flow and rising paper profits is why most stocker operators often go broke during high cattle prices but get rich during low cattle prices. Inventory paper profits only become real if the appreciated animal is sold and not replaced. Therefore, maximizing total profit requires a variable sized operation that can expand inventory during cheap cattle prices and liquidate inventory and shrink during periods of expensive cattle to "realize" paper inventory profits.

Cash flow is created by the sale of cattle. The more opportunities you have to sell and buy back cheaper cattle the better your cash flow will be. If opportunity costs are figured, in effect, you sell and buy your cattle every day whether you actually do it or not. Now here's the kicker.

In cash flow terms, it doesn't really matter what you sell your animals for. What matters is the amount of weight you put on, how cheaply you can put it on, and the difference between what you sell them for and what you can buy a replacement. You can actually have an increasing cash flow in a falling market if your replacements are falling faster than your sale animals.

The stocker business requires sharp pencils and calculators to determine which weight, quality, and even sex, offers the highest net cash flow. As we will see not all gain is equal in value. Too many of us have boxed ourselves into certain weights and grades that may not be the most profitable buy on a particular day.

Not All Gain Is Equal in Value

To illustrate this is a chart using 1986 prices (almost identical to 1996 prices) for cattle in the Southeast in which we have figured the value of each additional hundred pounds of gain with the exception of the 800 to 1100 lb weight range. The 1100 pound steer is a slaughter steer and the value per cwt is determined by dividing the dollar difference between the 800 lb steer and the 1100 pounder by three (300 lbs or 3 cwt).

Value of Gain Weight	Price/cwt	Price/cwt	Value of Gain per cwt
200	$95	$190	—
300	83	249	$59
400	74	296	47
500	62	310	14
600	58	348	38
700	55	385	37
800	53	424	39
1100	59	649	75

As you can see on this particular market, the value of each 100 pounds added is worth less until the animal is put on feed. The hundred pounds between 400 and 500 lbs is worth the least and the market only paid us $14 to produce it, whereas, it paid the cow-calf man $95 a hundred for the first 200 lbs.

The market is also telling us to own these animals all the way to slaughter, as well, as this gain has net value of $75 cwt even though the actual price per pound is only $59. Such situations often occur after a long downward price cycle that has wiped out the equity of highly leveraged speculators. As my Dad told me, a grazier always needs to stand as the feeder of last resort, because that is the point at which it will be most profitable.

Keep in mind that a forward margin, as this 800 to 1100 lb weight range illustrates, always creates a net value of gain higher than its nominal sales price. This is why you always want to seek out and take advantage of forward price margins. Conversely, negative price margins produce a net value of gain lower than the nominal sales price.

On this market, if a producer was selling a 600 lb steer and thinking of buying back a 200 pounder he could quickly figure that this would force him to swallow a $74 negative margin per head. To make a profit, he would have to produce gain for less than $39 cwt, inclusive of all costs.

I arrived at this figure by subtracting the cost of the 200 pounder and dividing it by four to represent the 400 lbs of gain. If we decided to buy a 400 pounder, we could quickly figure up that we must produce gain for less than $26 cwt to break even. This little exercise quickly illustrates that the traditional 400 lb stocker calf is the worst buy on that particular market. Also, if you will run this exercise on several different sell-buy combinations a stocker operator to be making money would need to be able to produce a pound of gain including all costs for less than $37 cwt to be profitable on this particular market.

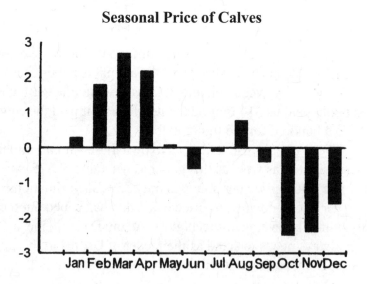

Seasonal Price of Calves

In addition to using the value of gain concept in risk management, it can also be useful in determining what weights to buy and what program to follow. Because our grass and overhead costs are largely fixed, our profits will be maximized by buying cattle that give us the highest value of gain. A commonly quoted figure for grass and animal costs is 20 cents a pound. On the market shown in the Value of Gain chart, let's see which weight of animal would give the highest value of gain. The results may surprise you.

Gain	Value of Gain
200-400	$53
200-500	$40
200-600	$40
200-700	$39
200-800	$39
300-500	$26
300-600	$33
300-700	$34
300-800	$35
400-600	$26
400-700	$30
400-800	$32
500-700	$38
500-800	$38

How many would have thought the $95 calves would have been the best buy? Or that the traditional 300 and 400 pound calves were the worst buys on that market? 200 pounds of gain on a 200 lb calf at 20 cents a pound cost of gain would have netted us $66 a head, but only $12 a head on a 300 or 400 pounder! As we can see from the above chart, the 100 pounds from 400 to 500 only costs $14 to buy, or less than our $20 production cost. Therefore, the 500 pounder is definitely the better buy from a net value of gain standpoint.

This extreme price discrimination between 400 and 500 lbs

is most often seen in the fall of the year in areas where stocker cattle are over-wintered on grass. The price premium for lighter calves is in recognition of the lower body maintenance requirement of the smaller calf. You can take three 200 lb calves through the winter on approximately the same amount of grass as one 600 pounder. This low body maintenance also produces higher gains per acre. (See Chapter 13 on grazing dairy calves for an example of this.)

In areas where calves are over-wintered on stored forages in a feedlot type situation, there is frequently a reverse price discrimination with the lighter weight cattle selling at a lower price per pound than a 500 pounder.

Note that once you cross the 500 lb threshold the value of gain stabilizes at almost a constant value for any additional gain. A major overlooked opportunity is the buying of feeder weight animals for exploiting short-term grazing opportunities.

A major problem most graziers face is that they can carry far more cattle in the spring than they can later in the summer. What works best from both a pasture grass control standpoint and a cash flow standpoint is to run a short-term set of cattle and a longer-term one.

Seasonal Prices of 750 lb Feeder Cattle and Corn

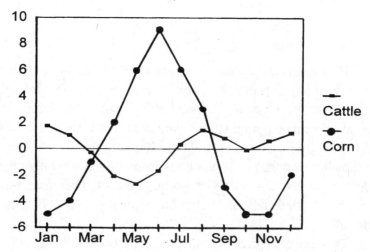

May is normally the lowest priced month of the year for feeder weight cattle (700 - 800 lbs) and August is normally the highest. This spread has to do with both feeder cattle availability (there are less available in August) but, more importantly, the price of corn (May and June are traditionally the highest corn price months).

Rather than scrapping and fighting over the handful of 450 to 500 lb stocker cattle that are available in the late spring, Northern graziers ought to consider going up in weight and buying the much more numerous and discounted 700 to 750 lb animals. A grazier has an "unreasonable advantage" bidding against a feedlot feeding the highest priced corn of the year versus little to no advantage against another grazier.

This weight animal should have no problem gaining over 100 lbs on a 45-day graze. Gains of cattle previously grazed on endophyte-infected fescue have been recorded as high as 6 lbs a day on spring planted annual ryegrass!

One of the graziers I always visit on my trips to New Zealand is bull-beef grazier Phil Taylor of Te Awamatu on the North Island. Phil primarily buys day-old male dairy calves and bottle feeds them up to grass. He buys these ultra-lites because his sale dollars buy the most replacement animals at that minimal weight category.

However, the other weight category Phil buys are 880 to 900 lb animals in the spring. These heavy-weights gobble up the excess spring grass and are sold at the end of the pasture surplus. Phil said that by buying heavy cattle he was able to avoid the extreme negative margins lighter weight cattle would have produced on such a short graze.

Phil said to keep in mind the spring pasture's surplus is not only wasted if not utilized, but will lower the quality of the whole pasture for the rest of the grazing season. The alternative is to mow the pastures, which is a sure dead-loss expense. Even if the animals just break even, Phil figures he is ahead financially because he has avoided the mowing expense.

Let's return to our value of gain analysis. Such analysis is most critical during the high priced phase of the cattle cycle when calves are priced at a considerable premium to feeder animals.

200 lbs of gain has been a fairly typical weight increase goal for many stocker graziers. Let's look at what a negative margin does to the value of gain.

600 lbs x .60 = $360
400 lbs x .70 = $280
margin $80
divided by 2 cwt = $40 cwt or 40 cents per pound of gain

As you can see, on a $60 cwt market the $10 cwt price rollback has actually lowered the value of the gain by $20 to $40 cwt. That's a little close for most of us. Is there anything we can do to create a higher value of gain in this market scenario? Yes, we can produce more weight per animal.

Taking the same price spread for illustration purposes, let's put on 400 lbs as Gordon Hazard recommends.

800 lbs x .60 = $480
400 lbs x .70 = $280
margin $200
divided by 4 cwt = $50 cwt or 50 cents per pound of gain.

Aha! As we put more weight per head on we not only make more money from production margin but our marketing margin rises as well. What if we put another hundred pounds on?

900 lbs x .60 = $540
400 lbs x .70 = $280
margin $260
divided by 5 cwt = $52 cwt or 52 cents per pound of gain.

Again our marketing margin increased! The more weight per head we have to spread the negative margin over the higher in value each pound becomes. What if we bought a lighter calf?

700 lbs x .60 = $420
300 lbs x .70 = $210
margin $210
divided by 4 cwt = $52.50 cwt or 52 and a half cents per pound of gain.

The lesson here is the lighter the animal the less impact

marketing margin plays. If we took this animal on up to 800 lbs with this same price margin rollback, we would increase our net value of gain to $54 cwt or 54 cents per pound.

Now what happens if the market for feeder steers goes up $10 cwt but the price of calves goes up $20 cwt? Let's look at the math.

800 lbs x 70 = $560
400 lbs x 90 = $360
margin $200
divided by 4 cwt = $50 cwt or 50 cents a pound.

We are now eating twice as much rollback but we are working on the same net margin as in our first example. In fact, we could pay a dollar a pound for a 300 lb calf, eat a $30 cwt rollback and still be working on virtually the same net margin ($52) as we were when we were only absorbing a $10 cwt rollback but selling our feeder animals for $60. This is why you see calves and yearlings come closer together in price as the feeder price level falls, and widen as the feeder price increases. The higher the price level the more negative margin you can handle.

Of course, the reverse is also true. What if we had to work on Argentine prices? Lets look at that.

800 lbs x 45 = $360
400 lbs x 40 = $160
margin $200
divided by 4 cwt $50 cwt or 50 cents a pound.

In other words, the Argentine grazier is working on the same value of gain as his North American counterpart who sells his steers for $15 to $25 cwt higher than he does. He does this by having a forward margin on his purchase rather than a negative margin as is commonly true in North America. In 1996, we saw calves go negative to yearlings. This produced the highest value of gain for the whole cattle cycle. In other words, the bottom is the top for stocker graziers.

This is the beauty of a margin operation. The nominal price really doesn't matter from a cash flow per head standpoint. However, it does matter from a return on investment standpoint.

Let's figure for that our gain costs us $25 cwt or $100 for 400 lbs of gain and it takes us a year to produce it. This means the American grazier selling his steers for $560 and paying $360 for them has a net per head of $100.

The Argentine grazier selling his for $360 or $200 a head less than the American is running on the same net $100 per head because he is only paying $160 for the calf. However, he is actually getting wealthy faster than the American because his return on investment is much higher.

A $100 return on a $160 investment in a calf is 62.5%.

A $100 return on a $360 investment in a calf is 28%.

In other words, the Argentine grazier is creating wealth twice as fast as the American on a far lower nominal market. What this shows us is that return on investment in a margin operation is inversely related to the price of your replacements.

Let's go back to our original figure of selling yearlings for $60 cwt or $480 and buying his replacement for $70 or $280. Remember, he has the same net margin per head of $100 but look at what happens to his return on investment when the cattle get cheaper. A $100 return on a $280 investment in a calf is 36%.

So for a margin player, cheap cattle create wealth faster than expensive cattle. This is why a variable size herd that can expand when cattle get cheap and shrink as cattle get expensive will not only allow you to pocket inventory profits but will produce a higher total return on investment as well.

Hazard Self-financed from Cash Flow

Since 1972, Gordon has been completely self-financed from his own cash flow. Gordon advises that a grazier should not routinely use leverage for the cattle portion of your operation unless your banker is very familiar with the cattle business. "You can't raise a family on a series of one-shot deals. It has to be structured as an on-going enterprise that can run on its own cash flow," Gordon said.

Even good credit borrowers can have their credit lines jerked in a falling cattle market. "If you want to borrow money, borrow it for the land," he advised. "Bankers understand land. Most don't understand the cattle business."

Gordon said he always tried to buy land with some salable size trees on it and often got enough out of the timber to pay his fence, water and grass planting expenses.

He said his total cost of gain is around 20 cents a pound. This includes receiving costs, labor, water reticulation, fertilizer, equipment rent, death loss and fence repair.

While he is no longer personally interested in getting any larger, he is helping all of his four children get into ranching. Their ranches are stocker operations financially structured like his.

"I think 1500 to 1800 steers is a nice size grass operation for one man. You can make an excellent living from it and it is still small enough to be hassle free," he said.

Pasture Profits

* Stocker operations don't go broke because of negative margins, they go broke because of lack of cash flow.

* Recognize and understand the difference between inventory profit and loss and net cash flow increase and decrease.

* Profits will be maximized by buying cattle that offer the highest value of gain.

* Not all gain is equal in value. For a healthy cash flow stay flexible regarding weight, quality and sex to buy as replacements.

* Grass growth can affect marketing flexibility by running a heavy short-term set of cattle for the spring lush along with a longer-term set.

Chapter 6

An End to the Crap Shoot, Using the Cattle Cycle to Create Wealth Faster

"Wealth from gambling quickly disappears; wealth from hard work grows." Proverbs 13:11

I f you will study the seasonal price charts for the "average" year previously shown you will see that the "average" range from the highest price to the lowest price is only around 3 to 5 percent. On a 70 cent market this is a total variation of around 2 to 3 cents between the high and the low. This is not enough difference to really matter to most people.

This seasonal demand variation should perhaps be considered as the surface froth atop a huge tidal wave. The huge wave is the cattle cycle. And, it very definitely matters.

The average cattle cycle is ten years long. This normally is made up of three years of high prices, two years of falling prices, three years of low prices, and two years of rising prices.

The cattle cycle is not something to be feared, but is for the flexible, forward thinking grazier a great opportunity. Playing the cattle cycle successfully starts with the realization that, unlike you, most producers are not paying attention to where the market is going, but only where it has been.

10 Year Average Fat Cattle Prices

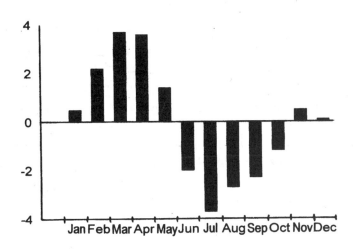

There are three profitable strategies in buying and selling livestock or any commodity. These are:

Buy low/sell high.

Sell low/buy back even lower.

Buy high/sell even higher.

However, as we shall see these three strategies differ in which one produces the most "real" profit or net cash flow.

Unlike the upside of the cattle market, which has a price trajectory similar to a rocket, the down market occurs in a long series of breaks and rallies, and is not a straight line decline.

Cattle feedyards make money two ways — on the cattle they feed for themselves, and on the cattle they feed for others. The capital requirements to fill a 100,000 head feedyard are mind-boggling. Modern feedyards were built to profit mainly from feeding cattle for others on a feed markup and yardage basis. Its profits are highest when its feedyards are full. To the feedyard owner the price of cattle has very little relevance to their profitability.

A feedyard's primary marketing job is to find people willing to bet on the cattle market. These people are primarily drawn from the same clientele who keep Las Vegas booming. Cattle feeding is a crapshoot and that's why they play.

Thanks to long-term relationships with banks and PCAs in their region, cattle feeders make it possible for their clients to heavily leverage most of the price of the feeder cattle and all of the feed. Because of the way these deals are structured, both real and paper profits can quickly be leveraged into more real cattle owner-ship. During profitable periods, this leverage works like super compound interest and allows the investor to buy many more cattle to put on feed than he is selling off of feed without increasing his initial capital input. This is called "playing on the house's money."

This is why the upside of the cattle market is so explosive and why feeder cattle tend to be quickly bid premium to fat cattle after even very short profit runs.

During the cattle market's slow downward slide, most investors don't get concerned as long as they are playing on the "house's money." Cattle feeding can run for relatively long periods at a loss and still pay premium prices for feeder cattle. Only when losses start eating into their "real" capital — their initial investment — do investors start leaving the game, often all at the same time. During such times the feeder cattle market can suffer a sharp price break due to a drop in the number of players. These breaks tend to be temporary and are usually corrected in the time it takes to slaughter one complete set of cattle on feed, or 90-100 days.

A true "wreck" or market bottom, on the other hand, is caused by the tremendous increase in the beef supply brought about by the liquidation and slaughter of the cow herd. Often this liquida-tion phase is worsened by drought and other weather difficulties that both increase the price of corn and diminish the supply of grass as happened in 1996.

A shift to the downward transition is forecast by a fat cattle market that never quite rallies to its previous high after it breaks. A feeder cattle market that starts to sniffle while many of the heifers are being routed into the herd will catch pneumonia when these heifers are rerouted back into the feeder cattle market.

As the general price level of fed cattle decreases, cattle

feeding becomes much more sensitive to the price of corn. At times, the feed cost to produce a pound of gain may actually exceed what it will bring, as it did in 1996. At these times feeder cattle will be negative to fats and feeders will seek to buy as heavy a feeder animal as possible to "run the grass forward" in price, as they say.

Choice grade steers hold their value much better during the down cycle than heifers, dairy breeds and high percentage Brahman-blooded cattle. Feeder cattle farther than 400 miles from Liberal, Kansas, tend to decline in price more than those closer to the High Plains, and seasonal basis swings can be extreme, particularly in the South. The reason for these steeper declines is the increase in feeder cattle supply allows cattle feeders to be increasingly choosy about the cattle they feed.

Given a choice, most feeders would rather feed steers than heifers, yearlings rather than calves, buy them closer than farther, and have them grade Choice rather than Select. For this reason, you must upgrade the quality of the calves you buy as prices fall.

As the price of fed beef gets closer in price to its cash cost of production the amount of negative margin rollback a cattle feeder can absorb is reduced and heavy yearling steers become the preferred class. In 1996, thanks to very high grain prices and very low fed cattle prices, the price structure became inverted with the very heavy feeder steers bringing a considerable premium over lighter weight feeders and calves. This was the absolute best of times for a stocker grazier. Therefore, you should always pray for very high grain prices.

The feedyards getting out of calf feeding is what drops the demand for calves and narrows the price margin between steer calves and steer yearlings. While the overall price of steer yearlings are coming down the premium for steer calves is falling faster. This narrowing margin actually increases the value of the gain and, therefore, net cash flow.

If you can stabilize your sales price by booking at least a portion of your cattle while your replacement costs are continuing to fall, the downward phase of the cattle cycle can be your most profitable net cash flow period. This is true even though you may be showing inventory valuation losses. This good fortune is magnified by the fact that your paper losses are tax-deductible.

Coming down the price curve is a nerve-wracking experience for your banker. He is primarily concerned about the value of his collateral, which he sees declining in value. For this reason it is best to have your banker out of the livestock leverage business before the downward transition. However, don't forget about him. Most bankers are willing to lend and lend heavily once they see feeder cattle prices hit bottom and stabilize.

If you are going to use leverage with livestock, the bottom of the market is the best time to do it. For example, during the summer of 1996 you could buy enough 350 lb calves to make an outgoing truckload of 800 lb yearlings for less than $7000. Your dollars buy a lot more livestock at the bottom of the market than they do at the top.

The upside of the cattle market is always a hundred times more explosive than the down side and a hundred times more dangerous. What makes it dangerous is the mistaking of inventory profits as real profits. Let's look at the three "profitable" strategies in terms of cash flow rather than paper profits.

Buy Low/Sell High

Inventory profits are always figured on a buy-sell basis. This means that you will show your highest profits during a rising price market. However, these profits are "paper" profits and are not real in an on-going operation because inventory sold has to be replaced to create future sales and cash flow. Inventory profits can only become real if the inventory sold is not replaced. Therefore, a buy low/sell high strategy necessitates a shrinking herd to produce "real" profits. Consequently, this strategy is best used as a "side deal" and not your main business. For example, some stocker graziers will buy a herd of cows at the bottom of the market, ride them up in price and cash them in at the top. However, young pairs (3 in 1's) bought near the bottom of the market offer a much better return than cows alone because you have three critters going up in price instead of one.

Rising Market

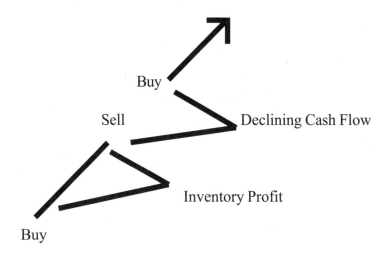

What you are going to do with these profits should be a major part of your planning process. Many choose to spend them on tax deductible ranch improvements and fertilizer that will allow the ranch to carry more cattle during the next falling price cycle. Sometimes a savings account is the best place to put your money.

There is a great temptation to invest in beef cows at the top of the cycle due to the high prices of weaned calves. Don't do this! A fully grown, fully priced animal at the top of the market can do nothing but lose value. For this reason fully grown animals are only profitable when bought at the bottom of the market or early in the market rise. Even more than stockers, cows must be liquidated at the top of the market because they are not growing in weight. The primary profit earner of a cow-calf operation is the sale of cull cows not calves.

This same temptation also exists with feeder heifers. If you get the urge to turn your heifers into cows remember this little truism. "No heifer calf born in a high market ever grew up to have a calf that was sold on the same high market."

The above truism about the short life of the high price phase of the market calls into question the wisdom of ever growing out replacement heifers for your own account. For others, to be sold at the top of the market, yes definitely! Buying and growing out replacement heifers is a viable, short-term enterprise and an excellent option when steer price rollbacks become extreme as the market peak draws near. Keep in mind this replacement market appears to defy the laws of gravity after the peak of the market and to stay higher than it should, longer than it should.

The only real price in the cattle business is the fat cattle price. This is the engine that pulls the rest of the cattle train. However, the cow herd will continue to liquidate or grow for several years after the fat cattle market has clearly turned the corner and is falling or rising. This market phenomenon is called lag. Markets always continue in a given direction long after the economic fundamentals that caused a particular circumstance have changed because people want to keep doing what they are doing. They will typically continue to do it until it becomes too painful to continue. Consequently, an aware grazier has to always stay fixated on the fat steer price. Everything else eventually comes into correlation with it.

Buy High/Sell Higher

A buy high/sell higher strategy works only as long as you can maintain a constant margin between your sale animal and its replacement. If your replacements are increasing at a faster rate than the price of your sale animals, you will have a decreasing net cash flow at the same time your tax liabilities from your "paper profits" are increasing.

To maintain a constant margin in a rising market requires the reverse buying strategy to the falling market. Rather than upgrade your purchases, you down grade them. In other words you switch to heifers, dairy steers and other discounted breeds. Whatever produces the maximum net cash flow on that day's market is the animal to buy. A rising market is also when it is most critical that you buy your replacements back as closely to when you sell them as possible. Any time delay in replacing inventory in a rising market will worsen net cash flow. Also, any attempt to expand numbers in a rising price but falling margin market is guaranteed to run you out of cash.

Rising Cattle Cycle

Rather than being the "best of times," fast rising markets are the "worst of times" on a cash flow basis for stocker graziers.

For example, in 1997, 450 lb calves increased in value by around $225 a head over the previous fall. This means your profit per head on your last set of calves sold would have had to equal this amount for you to be able to stay even in cattle inventory without borrowing money. Rather than making record profits most stocker operators were actually hemorrhaging cash at an extremely dangerous rate. This was made even worse by the huge tax payments they owed on their past inventory "profits" and by the fact that they many were leveraging these "profits" into an expanded inventory.

It is very important that you understand the difference between "inventory profits" and "cash flow."

Peter Drucker calls inventory profits "taxable accounting fiction."

The operative word here is taxable.

In a rising market, when inventory replacement costs are requiring maximum cash reinvestment, the government is taking 30 percent of it away in taxes. Drucker said that most accountants keep your books for the Internal Revenue Service, not for you. These records are at best a historical account of what happened, but are of little use to determine what is going to happen. What you need is a firm grasp on what is happening to your cash flow. Drucker said that running out of cash was the number one reason for bankruptcy.

Ironically, the "best of times" cash-flow-wise for stocker operators is not a rising market but a falling market.

By booking at least a portion of your purchased animals, your sales price is fixed but your replacement costs continue to fall. This can result in your highest net cash flow per head of the cattle cycle. It is also possible to at times show an "inventory loss" while doing this and avoid taxes as well.

Conversely, the worst time of the cattle cycle for a stocker operator from a cash flow standpoint is the upward transition phase. This is both because of the tax problems outlined above and because replacement costs tend to be bid up faster than the price of sale animals. This is the result of new players entering the game attracted by the noise of "record profits" and the old players expanding numbers to chase these same "fictional" inventory profits. Keep in mind the only way for an inventory profit to become "real" is if it is not replaced. However, if you do this you are out of business.

For operators trying to operate solely from their own cash flow, expanding numbers in a rising market is a sure recipe for bankruptcy. Put a pencil to it.

When the increase in replacement cash costs exceeds the net cash margin of the out-going set of cattle, the only way you can stay even cash flow-wise is to reduce numbers, buy lighter cattle, sell them heavier, downgrade in quality, change to heifers, dairy steers, or a combination of all of the above. You cannot afford to replace like for like, or number for number.

Keep in mind that in a margin game all the "real" profits are generated at the time of an animal's purchase, not at the time it is sold. If a sell-buy inventory swap won't pencil out on the front end of the deal, don't do it.

One of the things you cannot allow to happen in a rising market is to allow a significant break in time between when your out-going animals are sold and you buy your replacements. For example, calves sold in the spring of 1997 for what 800 lb yearlings sold for in the fall of 1996. Ouch!

Booking programs, which are so smart in a falling market, kill you in a rising market unless your replacements are bought at the same time.

In the reverse of the falling market, if you fix your sales price while your replacement inventory is rising you guarantee yourself one major cash-flow crash.

For graziers without year-round forage resources, a rising market is the best time in the cycle to own animals all the way to slaughter. Not only will this help prevent a break in time but will maximize gross dollars per head.

Declining Market

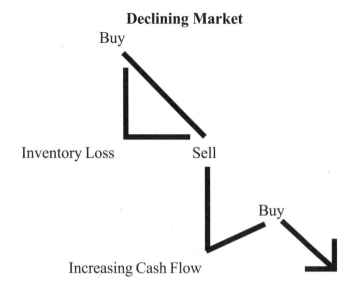

Buy

Inventory Loss   Sell

Buy

Increasing Cash Flow

Sell Low/Buy Back Even Lower

Interestingly, of the three strategies it is the normally over-looked strategy of sell low/ buy back even lower that has the greatest opportunity for producing the most net cash flow. As we've previously discussed a buy/book program whereby your outgoing price is fixed but your replacement costs continue to fall produces the most net cash flow.

The stocker business is not a game of absolute prices but one of margin. To a stocker grazier the good times are the bad times. He has to learn to see the high priced period as the bottom of his profit cycle and the low priced period as the top.

Following this strategy requires two things — one, the unshakable belief that a depressed market will always recover, and, two, that a soaring market will always break. The cattle cycle has been one of the most predictable economic cycles for over 100 years. Use it!

For whatever reason, Argentines do not experience the price volatility that we see in the USA and this tends to make ranching a more stable, predictable business in Argentina.

In 1996 Argentine beef cow prices were around $250 per head, or approximately the same as in the USA. However, this was down only $80 from the peak price several years earlier. In the USA the price was $600 to $700 off its peak from three years earlier. 1996 Argentine slaughter steer prices were $35 cwt versus $42 cwt at the peak. This was a drop of only $70 a head, whereas in the USA prices fell by over $200 a head.

Thanks to this minimal negative margin over the course of the cattle cycle and the low costs of Argentina's all-grass production system, Argentines are able to pretty consistently net 15% to 30% on capital invested in ranching. There are few business alternatives that can do that in Argentina. As a result cows tend not to be liquidated during times of low prices but "repositioned" on cheaper, "less favored" land.

Beef cows are best suited to hot, wet climates. Over 70% of the world's beef cows live in tropic and subtropic regions. In Argentina, the beef cows are increasingly being shifted to the North of the country where the climate is similar to Louisiana and Florida and unsuited for temperate grasses and grain production. The temperate finishing zone is becoming almost exclusively a grain production area and a grass/growing finishing zone for weaned calves from the north. In this area, the growing and finishing of weaned calves has traditionally been the centerpiece activity and beef cow-calf and grain production the ancillaries.

Argentines traditionally pay in commodities rather than cash for inputs. For example, ranch leases stipulate so many pounds of beef and crop leases, so many cwts of grain rather than a cash price. By using commodities rather than cash, ranchers' inputs rise and fall in concert with the price of the commodity produced. This prevents the cost/price squeeze that creates so much pain for North American farmers. Compared to the USA, Argentine ranches tend to have many more cattle and tend to be vertically integrated from birth to slaughter (although not necessarily at the same location). Argentines figure 1000 beef cows are the minimum needed to employ a full-time cowboy. When all costs (house, horse, taxes, etc.) are added up, Argentines find one cowboy costs around $3000 a month to employ.

Argentina is more costly in almost all respects than the USA including, surprisingly enough, food costs. Beef in the supermarkets in the USA is considerably less than in Buenos Aires even though finished steers in the USA average $20 cwt higher than Argentine steers.

The traditional rule of thumb for ranch land ownership in Argentina has been that the land price should not exceed the value of the animals grazing it. However, today this is seldom the case. Argentina's land has been priced at $1500 to $3600 a hectare ($600-$1500 per acre) and the land is now often twice the value of the animals. As a result, leasing land is becoming more common.

With weaners (stocker cattle), the animals are weighed on and off the land and the landowner traditionally receives 60% of the animals' gain and the cattle owner 40%. All cash inputs (medicine, vet, etc.)are split 50/50 between the land owner and the rancher. The exception to this formula are Holstein steers.

With Holsteins the landowner gets 70% of the gain in acknowledgment of the high cost of overwintering the heavy, slow finishing Holsteins. (Holsteins are exported to Europe as finished Euro-beef, not to North America as hamburger.) With cow-calf, the landowner gets 55% of the calves (by weight) and the rancher the rest.

With crops, the rate is adjusted (by competition) in response to the price of grain. For example, 35% of the production for the landowner was common until recently. In 1996 grain was up to 45% to 50% and some landowners were demanding a cash down payment equal to 10% of the value of the land ($80 to $90 per acre) before planting.

Adding Value to a Resource Creates Real Wealth

Sustainable wealth is created by adding value to a resource, in our case, turning "worthless" grass into dense, high value, edible human protein. So, why does price speculation get so much attention? Here's a theory.

There is a tendency for all of us to want to know what the score is in any game we're watching. The more the media can couch business and economic stories in the sports vernacular of scoring, winning and losing, the more interested we become in reading about it. "Winning" in commodities has become symbolized by a rising price, and "losing" by a falling price. That there is a winner for every loser due to the necessity of offsetting positions is ignored, because it would destroy the sports metaphor. Also, since there can be no actual winners and losers in a game that never ends, the media takes upon itself the duty of declaring daily winners and losers to keep the game interesting for the spectators.

By periodically screaming "the sky is falling," as the media did with the October 1987 stock market "crash," they can send millions who do not follow the daily game scrambling to buy a newspaper or turn on the TV. Bad news is always better than good news in drawing a crowd and so gets played up stronger. Don't get suckered into this media circus.

By continually watching sell-buy margins rather than prices, you can easily pick up on these signals of when to expand, stand pat or liquidate. In general, a widening sell-buy margin means a safe time to expand even through leverage. A narrowing sell-buy margin signals to stop expansion and consider a partial liquidation if you have accumulated significant unrealized inventory profits.

Few Bankers Understand Stocker Business

Bankers are trained to lend based upon collateral and stable absolute prices. They prefer buy-sell accounting with a hedge or booking program because they are first and foremost interested in getting their capital back. While this is an excellent strategy in a falling market, it absolutely kills you in a rising market by increasing the gap between your sale animal and its replacement.

Bankers will readily tell you that they are short-term lenders and are not in it for the long run and only consider one deal at a time. Can you build a livelihood, plan to send your children to college, even feel financially secure on a series of "deals"? Bankers have a place in our business, but not to the point that our very financial life and death are dependent upon them. They are best kept to what Louisianians call the lagniappe — the side deals.

For example, in the spring of the year many of us can handle more cattle and like to "load-up" our pastures for the 60-day spring lush. In many climates this serves as a good time to "finish" heavier cattle that will be routed to the feedlot at the end of the lush, and to "start" calves that will be grazed through the summer and fall. These are called "loadup calves" in the South. However, this strategy while excellent in increasing the efficiency of our grass harvest requires us to have two sets of cattle under our ownership at one time, and a corresponding short-term need for more capital.

A good program to work on with your banker would be to book your heavy outgoing cattle for a late spring sale and borrow against that fixed price to buy a spring/summer set of lighter cattle. Almost any banker will lend you money for the loadup calves for 90 days with only the booked set of cattle as collateral. This is called a self-liquidating loan and from the banker's standpoint is the best way to use his bank's money.

The Bought-Weight Hedge

Here is a concept you may want to explain to your banker that will give his capital risk protection but not totally kill you in a rising market. This is a program called the "bought-weight hedge."

Within any purchased animal is a portion that we purchased, "the bought-weight," and a portion that we created from low-cost grass. Of course, the portion we grow is much lower in cost than the portion we buy. The risk to the bank's investment is that the value of the bought-weight portion will deteriorate to the point that we are unable to repay the loan. This is a real risk. In the early stages of the downward transition, yearlings sometimes bring less total dollars than the original calf cost. However, we can manage this risk by booking or hedging only a weight equivalent equal to the bought-weight at a price equal to that required to protect the banker's portion of the loan. Most stocker cattle loans are made up from 75 to 85% of the bank's money and the remainder being our equity contribution. All the bank is concerned about is getting their money back with interest.

Here's a feedlot example. We have a pen of 200 steers that were priced at 750 lbs at 60 cents a pound. This is an on feed weight of 150,000 lbs. We plan on selling a pen of 1100 lb steers or 220,000 lbs at an unknown price, but we are betting it will be higher or we shouldn't be in a feedlot. If we can book or hedge 150,000 lbs at 60 cents we can offset most of the risk and still have 70,000 lbs to gamble with on a price increase. Since fat-cattle futures contracts are in load-lots of 40,000 lbs, four contracts would offer us price protection on our high-risk bought weight. What can be really sweet is if we can hedge or book a forward margin or price increase in for our bought-weight on the front end. It doesn't happen often but it does happen.

With the exception of the upward transition period, making a routine practice of booking a portion of your weight for future delivery will, on average, result in a higher average value of gain and an increased net cash flow.

Retained Ownership

If feeder cattle are priced premium to fat cattle, there is very little incentive for a grazier retaining ownership except as a way to overwinter his equity and maintain a sell-buy relationship in the upward transition phase. Of course, the heavier your feeder cattle become the more sense it makes to retain ownership. Many people figure 850 lbs is the balance point. If they weigh less than 850 lbs sell them as feeders and if they weigh more than 850 lbs feed them. However, in the spring of 1989, even 1000 lb feeders were still priced two dollars over slaughter cattle, so there are exceptions to this rule.

It appears from market analysis that the overwintering of cattle in the feedlot sure works better than grain feeding them when the grass is green. According to cattle analyst Ed Uvacek, from 1978 to 1988, cattle killed in April made money six times out of ten. Those killed in March, May, and June made money five times out of ten. But those fed through the summer and killed in August or September only made money once in ten years. (See Fat Cattle chart on page 55.)

No doubt much of the problem with finishing cattle in the summer is that this is the time of the year when grain prices traditionally peak. For example, between 1987 and 1996, late May corn prices averaged 10% higher than January. It is this other dimension of an ever-changing grain and gain price that many graziers fail to take into account when retaining ownership of their cattle. If you are planning to have cattle fed in the spring and early summer period, some sort of protection from an unexpected rise in grain prices could be money well spent. Some graziers will pre-book an amount of feed for a fixed price. Others will use the futures market as a hedge. Most custom feeders will have some sort of grain price protection program, but you usually have to ask for it.

Quite a few Southern graziers who formerly summer feedlotted their winter-grazed stockers, now continue to graze them through the summer to avoid the heat, high grain prices, and the often very wide Choice-Select spread that mars the summer finishing of cattle. Many of these cattle do not go on feed until the fall when they will weigh 1000 lbs or more.

Many Midwestern farmer-feeders will buy 600-800 lb cattle in the South in early May, and stop them off for a cheap extra 150-200 lbs of grass gain in the Flint Hills before putting them on feed. This program works really well with fescue-grazed cattle as they have a lot of compensatory gain built up in them. A great many cattle are put on feed at 1000 lbs in late summer in the Midwest.

Use the Feedlot as a Banker

Retaining ownership through the feedlot is a good idea during the upward transition phase. Of course, this breaks your cash flow and disrupts your sell-buy replacement program. Having all of your equity tied up in the feedlot is particularly aggravating with a spring set of cattle and a long green grass season still ahead.

Many of the larger custom feeders understand this problem and will lend you back all but a hundred dollars of your incoming feeder's price to encourage you to feed with them. This will allow you to buy your replacement calves as usual. This type of arrangement will allow you to ride out a market collapse in the feedlot but still buy your replacements on the low market.

Cattle leverage should be reserved for extraordinary, short-term events. There is no way for you to eat well and feed a banker at the same time.

Placing Bets on an Uncertain Future

In theory, the futures market appears to be a good way to offset risk. In reality, I want to warn you that it takes an extremely strong person not to be sucked in and consumed by it. When the market is going your way, the quick and easy money that piles up in your account can be as heady and euphoric as the most potent drug and just as addictive. I have seen many farmers and ranchers totally lose interest in the real world of cattle ranching and spend day after day mesmerized by the blinking numbers on the toteboard screen. There are times in our lives when the only alternative left is to take a strong, potentially addictive drug. However, I urge you to explore every other risk management alternative before you seriously consider using the commodity markets.

The committing of capital based upon an uncertain future is a major source of risk. In the cattle business, we have a long slow production cycle that allows us to predict the future far more accurately than can be done with short-cycle commodities. However, what can't be predicted are things that can impact the price of beef from outside the cattle business.

Few American beef producers were paying attention to the government's dairy buy-out program in the mid-1980s and few British beef producers had even heard about BSE until the storm broke around them. Graziers is East Texas had not seen drought like the one in 1996 in over 100 years. No one could remember snow drifts 20 feet high like those seen in North Dakota in 1997 or could foresee the floods they would bring when they melted. Planning for such unforseen events is part of your job as a grazier.

The Arabs say people who try to predict the future are either insane or irreligious. Peter Schwartz, the author of **The Art of the Long View**, said it is the planning for only **one** future that is insane.

He said strategic decisions (what, where and when to produce a particular commodity) must be sound for all plausible future scenarios in order to be financially prudent. Most people make bets on the future based upon the continuation of the present situation. Few have a fall-back or contingency plan if this situation should reverse itself. Probably even fewer have a plan for what to do if they should suddenly find themselves in that proverbial "right place at the right time."

Schwartz said that future planning should always include at least three scenarios about the future. These are:

The official version of the future.

Our worst nightmare.

The best of all possible worlds.

He said the event we must always plan for is the one we don't think will happen. These plans should always be in writing to be of any use. Few of us can think straight in a high pressure, crisis situation.

While planning for the worst is the most important scenario, he said it is important not to dwell on the negative. Allow yourself to dream up a scenario that would give you all that your heart desires. Also, never end a negative scenario until you have figured a way of escape and recovery.

Pilots are taught to mentally think through all sorts of equipment failures and weather emergencies that might befall them. Mentally rehearsing successful recoveries from near disaster creates "action paths" in the brain. These paths allow pilots to handle emergencies on "automatic" — without conscious thought. They are warned to never visualize the plane crashing. "If you don't want to go there, don't allow your mind to go there," they are warned. Dwelling on negative outcomes, almost invariably produces them. (Try to find what opportunity door would be opened by what appears to be an unmitigated disaster.)

Schwartz said that it was very important that to always have a good sense of where you are in the market cycle. What you are always looking for are "surprises" that do not fit within the official future. Be attuned to that lone voice crying in the wilderness. He said widely reported predictions seldom occur because people react to them and often create the exact opposite of the prediction.

He said all economic cycles tended to be emotionally mis-
leading because things tend to look their best right before they
collapse. This anomaly is very difficult to comprehend unless you
have lived through at least one previous cycle. The people most
interested in not making future mistakes are people who have made
mistakes in the past. He described this as the primary advantage
older people have over younger people. The "old hands" know the
way it is, is not the way it is going to be.

Pasture Profits
* As prices fall, upgrade the quality of the calves you buy.
* Your most profitable net cash flow period can be the downward
phase of the cattle cycle.
* The bottom of the market is the best time to use leverage with
livestock.
* To create future sales and cash flow, inventory sold has to be
replaced.
* The sale of cull cows, not calves, is the primary profit earner in a
cow-calf operation.
* When steer price rollbacks become extreme (a sign of the market
getting near the peak), buying and growing out replacement heifers
is an excellent option for a short-term enterprise.
* Graziers who are looking ahead will stay fixated on the fat steer
price. Everything else eventually becomes correlated with it.
* When the sell-buy margin narrows, this is a signal to stop expanding. Consider a partial liquidation if you have accumulated
significant unrealized inventory profits.
* For stocker graziers, the best of times are the bad times for
everyone else. Learn to recognize that high prices are the bottom of
a stocker's profit cycle, and low prices are the top of the stocker
profit cycle.
* Remember, a depressed market always recovers, and a soaring
market always breaks.
* Problems can be buying opportunities. Be prepared to take
advantage of droughts, floods, blizzards, high grain prices, low
cattle prices.

Chapter 7

Partnering Pays

"A workman may eat from the orchard he tends, anyone should be rewarded for protecting another's interests." Proverbs 27:18

A lot of the winter pastures in the Gulf Coast area are annual ryegrass flown into soybeans by airplane in early October. Most of these pastures are ready for grazing by mid-November after the soybeans are combined. As in the wheat country of the High Plains, the person who is the grazier is usually not the farmer growing the beans.

Most graziers and farmers have found it better to specialize in one or the other and lease and sublease ground from each other. Some have formed loose "partnerships" with the lease based on various kinds of crop share arrangements to avoid cash rents.

The big feedyards in the West have long run winter cattle down here, but the fastest growing group of clients are Midwestern graziers who use our southern winter pastures to get a head start on the season. A 250 lb calf will gain around 300 lbs during the winter and be a prime 550 lb yearling ready to tear it up on the Midwest's summer pastures in May and will be "cheapened back" considerably, as they say.

More and more graziers are "partnering" with a fellow

grazier in the Midwest by selling him gain in the winter and buying it in the summer. This is particularly true in years when soybean prices are high and there is good demand for winter pastureland from bean farmers.

For several years, Buddy Moody of Poplarville, Mississippi, and Bob Skinner of Lexington, Nebraska, have partnered in this manner on stocker cattle. The cattle graze on Moody's ryegrass in the winter and on Skinner's native grass in the summer.

This partnership allows Moody to lease out his land for summer crop production and has allowed Skinner to avoid buying cattle during the traditional spring "grass fever."

The "partner" cattle are traditionally bought in January and are grazed in Mississippi until the first week of May, and in Nebraska until the first of September.

Alabama grazier, Pete Reynolds, partners with Florida cow-calf producer, Norman Stokes by buying some of Stokes' calves and grazing some of them for Stokes on a gain basis. Frequently Reynolds will then turn around and partner in the feedlot with a corn farmer a portion of the cattle he purchased. Partnering can both lower the capital needed to harvest your grass and spread the market risk.

Pasture Gain Prices

Winter pasture gain prices are dependent upon the price of corn and the price of cattle. Over the last ten years they have ranged from 25 to 42 cents a pound. As this is being written, gain prices are around 32 cents a pound.

Is there any money for the grazier selling gain for 32 cents a pound? While the results are frequently variable due to the severity of the winter, they are almost always positive.

I am going to use figures from the Rosepine Research Station's rotational grazing trial on ryegrass and clover that we reported in the June 1988 issue of the **Stockman Grass Farmer** to illustrate the potential.

The Rosepine station reported a gain of 710 lbs per acre from a six month ryegrass, crimson and white clover mix. 710 lbs x

32 cents is a gross of $227 an acre. Rosepine reported expenses of $84 for site prep, seeds and fertilizer. This leaves a return of $143 per acre. Not bad for land that can be leased for a dollar a month for the winter season from local soybean farmers.

Selling Gain Is Profitable

Gain 710 lbs. x .32 = $227
Pasture expense   - $84
Gross Margin per acre $143

The above example shows how the grazier's client can make out buying gain at 32 cents a pound.

While such returns are realistic and obtainable (the Brown Loam Station near Jackson, Mississippi regularly produces 900 lbs of gain per acre, although with a heavier nitrogen input and cost than Rosepine uses), I should warn you that the pasturage business has a terrible reputation for not living up to its promise for either the grazier or the client. Most of this disappointment is due to low average daily gains on the part of the cattle and this is due primarily to two things.

The Promise of Pasture Profits

One, the cattle are invariably turned out before the grass has had time to grow to the six- to eight-inch height necessary to carry the cattle through the winter at a good rate of gain. And two, the fixed price per pound of gain encourages overstocking. A high rate of gain per animal and a high rate of gain per acre are the opposite ends of a teeter-totter. If one goes up, the other goes down. The grazier soon learns that he maximizes his return by producing the most gain per acre at the lowest input cost in fertilizer and feed. This is not bad management if the client is a grazier who plans to take these cattle on to his own pasture and use them. In fact, this is the way he would want these animals grazed. The man who gets killed in the scenario is the one who was planning to sell them as feeder cattle and who needs a high gain per head.

What is needed in the latter scenario is a fee-schedule that would pay a higher price for each subsequent pound or fraction of a pound of average daily gain. Many sophisticated investors insist on such performance bonus contracts.

A regular weighing of the cattle would also help both the grazier and the client know what is going on with the cattle and prevent end-of-the-season surprises. Usually poor animal performance is as much a shock to the grazier as it is to the investor.

To encourage this practice, the client could pay a portion of his bill with each weighing. With today's low cost portable scales, weighing can be done with a minimum of fuss and shrink.

As more and more graziers become proficient at manipulating their grass so as to produce an intended animal response, and as the fee structure is liberalized to reward grazier competence, pasturing cattle can become a real profession with certifiable standards and an excellent reputation.

Joint Ventures

A joint venture agreement is a legal vehicle that allows two or more individuals to "partner" their assets and skills without incurring the rather complicated legal and tax implications of a true partnership.

The life of a joint venture is established at the outset and the relationship will end at that point. If the two individuals decide to continue working together, they can then enter into another agreement. If not, they shake hands and go on with their lives. Since all assets are held at the individual level and only the gross income produced by the joint venture is shared, such divorces are quick and painless.

All expenses incurred in the joint venture are deducted at the individual level. The joint venture does not have to file a tax return. Depreciation schedules on equipment, buildings and livestock are continued by the owner of these assets. Both parties in a joint venture are considered to be self-employed. As such, any draws from the joint venture are not expenses of the business and may not be treated as deductible expenses for income tax purposes.

A joint venture participant is only liable for expenses in the same percentage as the agreement spells out. For example, in a 50/50 joint venture the maximum liability any one partner can incur is 50 percent, whereas, in a true partnership both partners are fully liable for the debts incurred by the other.

A joint venture differs from a crop share arrangement in that all expenses are deducted prior to an income split. This prevents one participant from bearing all the production costs and risk.

Learn Graziering with Joint Ventures

Joint ventures are an excellent way to gain the benefits of specialized knowledge and skill that you do not have and would like to learn firsthand. Many graziers prefer to have a joint venture agreement with a custom feedyard rather than a straight feed markup. Most people feel they will get better service and attention if they have a "partner" who knows he will share in the profits or losses on the cattle.

A joint venture is an excellent way to see all of the costs involved in a new venture as a gross margin analysis is provided for you before each gross income split. Joint ventures prevent the replication of capital investments and allow both parties to more fully utilize existing investments and expertise.

Many ranches are too small in both land and animals to maximize the efficiencies available from management-intensive grazing today. A joint venture would allow several ranchers to pool their cattle, land and equipment into a more efficient size that would increase the profits of all participants.

In such a pooled arrangement, each contributor of cattle would make sure his contributed cattle are tagged or branded in a unique way. The progeny produced under the joint venture would be branded with a special joint venture brand. Upon dissolution of the joint venture, each party would receive his original cattle back. If any contributed cattle had died or been stolen during the joint venture, these missing cattle would be replaced from the stock of joint venture progeny and then the remainder would be divided by the percentage outlined in the joint venture agreement. An alternative would be to sell all the cattle and divide the monies by the agreed joint venture percentage.

Operating capital is lent to the joint venture by the partners in the same ratio as the agreed income split. At the end of the joint venture, the initial operating loans are repaid in full before the net income is split.

If one participant's cash drawings from a joint venture should exceed the profit percentage allowed to him or generated by the venture, he must make a personal note to the other participant for the debt incurred. In a joint venture, the balance outstanding must equal zero at the termination of the agreement.

A joint venture cannot use profits generated to purchase assets. All assets must be owned by individuals. Individuals may borrow money from the joint venture to purchase assets to be used in the operation, but care must be taken to make it clear that the asset is individually owned to prevent a possible ruling that you are operating as a legal partnership.

What a Joint Venture Should Contain

A joint venture agreement should be in writing and should:
1. Identify the parties involved.
2. State the purpose of the joint venture.
3. Clearly define the initial contributions of money, property, effort or other assets of each party and the agreed upon allocation of profits and losses among the parties.
4. Clearly define the allocations procedure and the responsibilities of each party under the agreement.
5. Joint ventures must be of a specified duration. This length of time must be established.
6. The distribution formula upon dissolution of any jointly owned assets must be defined and understood.
7. Clearly establish the sharing of management responsibility.
8. A provision should be included to negate all implications that a partnership exists, including a written provision that the joint venture will not be liable for debts and obligations of the other without written consent.

9. Clearly outline any buy-sell agreements or conditions upon which such agreements or conditions will be triggered for any jointly owned assets.

There may be some tax benefits involved in a joint venture for you. Discuss this with your accountant.

Specialization Increases Productivity

For several years, California grazier Harold Hunt has been in a joint venture with Al Christy, a local cattle order-buyer. This joint venture is a legal business arrangement whereby profits (and losses) are split after all direct expenses are paid. For Harold and his partner daughter, Pam, it has been a business arrangement made in heaven.

"A joint venture allows each partner to specialize in what they are good at, " Hunt explained. "I have no inclination to sit in a cattle sale every day and buy cattle. I'm a grazier and like the grass end of the business. Al, on the other hand, actually likes the auction circuit and has a good eye for cattle that will upgrade on grass."

The Hunts manage 350 acres of irrigated perennial ryegrass in the tabletop flat, Arcata "bottoms" on the north Pacific coast of California. The "bottoms" is the delta formed by the Mad River as it enters the Pacific Ocean. Naomi Hunt said they enjoyed the foggy, chilly summer climate of the narrow strip of land between the Coastal mountain range and the ocean. Land Naomi describes as "beyond the Redwood curtain."

Harold said he and Christy specialize in "upgrading" cattle and buy mostly from the bottom end of the cattle market. They buy stocker calves, trader cows, thin cows, pairs, anything they can put weight on or upgrade in price. "We found the bottom end of the market usually produces a bigger margin per head, which is important when you're splitting income," Harold said.

The secret to a good joint venture relationship is for the two parties to both trust each other and to need each other. "We had something he needed and he had something we needed. It's been a very good fit," Harold said.

Pam said Christy had a very good eye for identifying sick

cattle early and had taught her how to do this. "I really like working with Al. I've learned a lot," she said.

Pam worked office jobs for four years before getting fed up with it and going to work full time for her father as the head (and only) cowboy. She said working with animal behaviorist, Bud Williams, and learning how to work stock with a minimum of effort has made a big difference in both the health of the cattle and her enjoyment of the job.

"Our shipping fever problems are way down since Bud showed us how to take the stress off our calves," she said. Pam now does all the stock handling alone, whereas it once took four to five cowboys to gather the cattle for shipping.

The pride and joy of the Hunt operation is their 90-acre deeded homeplace. Hunt said a few years ago, he took out all the existing fences and subdivided the farm into 18 five-acre permanent paddocks with electric fence. These paddocks can then be split in half with temporary fence to increase stock density and/or lengthen the rotation.

Hunt has a long laundry list of benefits this intensive subdivision has brought him.

"We used to have to mow the whole place twice a year for thistles and weeds, but since we subdivided the place we have not had to clip once.

"We've been able to cut our fertilizer bill to 80 percent of what it formerly was.

"We've been able to cut our irrigation by 40 percent.

"We used to have lots of bloat, but once we started rotating, the bloat essentially stopped.

"Our internal parasitism problems are a lot less than what they were.

"The whole system just operates better. The cattle are much easier to get up and move. Subdivision is a big labor saver."

Hunt said the homeplace was used primarily for "finishing" stocker cattle prior to being sold with leased land used for the growing and framing. Al Christy buys a few head every week, and a highly variable stocking rate helps keep pasture utilization high.

Hunt might graze only 50 to 60 head on the homeplace in the dead of winter but go up to 200 in the spring and sell down to 150 for the summer. "Flexibility is the key to maximizing pasture utilization," he said.

Most of the cattle marketing is done in the late summer and fall.

"We use cows primarily for pasture conditioning purposes," Pam said. "We let the quality of the pasture decide if it gets cows or stocker calves."

Hunt doesn't like to own cows too long as their feet go bad in the soft soil and the high quality pastures over fatten them and predispose them to grass tetany problems.

Harold and Naomi Hunt's home sits in the middle of their farm. This allows Harold to sit in his hot tub on the house's deck and nurse a bad back but still watch the cattle graze with a powerful set of binoculars. Often he will see something he thinks needs immediate attention and will call Pam on the radiophone to take care of it.

"There's something so California about being given orders to go do hot, dirty work by a boss sitting in a hot tub," Pam said.

Pasture Profits
* Specialization promotes competence.
* Partnering allows both specialization and integration.
* Joint ventures are an excellent way to gain specialized knowledge and skills you don't yet have.
* Joint ventures reward competence.
* Flexibility is the key to maximizing pasture utilization.

Chapter 8

The Lure of Cattle Leverage

"The evil man gets rich for the moment, but the good man's reward lasts forever." Proverbs 11:18

A Midwestern reader once wrote to tell me that by his calculations the return to the investor in a stocker gain deal appeared far better than anything available on Wall Street. He wondered why the cattle industry was having problems raising capital. Let's take a look at what he was talking about.

Let's say 500 lb steers are about $95 a cwt and 800 lb are at $80 cwt. This gives you a net value of gain around $55 cwt.

If you are buying gain for $30 cwt this gives you a gross margin per cwt of $25. $25 times 3 cwt gives you a net margin per head of $75.

At an average daily gain of around 1.5 lbs per day it will take around six months to put on 300 lbs.

A $75 profit on an investment of $475 (500 lbs x $95) is a return of 16% for six months. Annualized this would be a return of 32%.

However, to really sweeten the deal try a 75/25 leverage ratio. At this ratio, you only have $118.75 of your equity invested. The bank puts up the rest.

The bank charges you $21.38 rent per steer for its money for six months, which you have to deduct out of your $75 gross margin. This leaves you a net of $55 on an investment of $118.75 for a return of 46% in six months or 92% a year.

In cowboy English, the above program allows a leveraged investor to nearly double his money in a year!

Non Leveraged Cattle

800 lbs at $80 cwt = $640
500 lbs at $95 cwt   - $475
Per Head Margin $165
divided by 3 cwt
Value per cwt of gain $55
Cost per cwt of gain $30
Gross margin per cwt $25
x three cwt. equals
a profit per head of $75   in six months
* $75 return on $475 is an annualized return of 32%

Leveraged Cattle

800 lbs at $80 cwt = $640
500 lbs at $95 cwt   - $475
Per head margin $165
Value per cwt of gain $55
Cost per cwt. of gain $30
Gross Margin $75
Interest at 10% $21
Profit in 6 months $54
Equity in steer $118.75
* Annualized return on equity 92%

Note, the investor has his total investment in livestock. He has no investment in land, tractors, horses and cowboys. The main reason we aren't rich is that we have too much invested in land and machinery and not enough in livestock. There are two ways to redress this — get more livestock, or get rid of non-livestock overhead and land. The first option is difficult because it requires even more capital. The second option is very painful because it forces us to admit to past mistakes.

Thanks to T. Boone Pickens and the corporate raiders, public company managers have discovered the quickest and surest way to get return on investment up is to get capital investment down. The same is true in ranching.

The Sale-Leaseback. One way to free up capital trapped in land is through a sale-leaseback whereby you sell your farm to an insurance company or investor on a guaranteed leaseback program.

The Rollover. The other technique is called the rollover. If you live in an area that is urbanizing, your land may have appreciated to the point that you can sell your farm, move to a more rural area, buy the same size farm but now have plenty of operating capital. The big trap to avoid is selling out and putting all of the proceeds into more land even though some state tax laws encourage this. Often this is just getting out of a small frying pan and getting into a bigger one. It is far safer to pay the tax and build liquidity.

The Garage Sale. Dr. John Rutledge of California's Clermont Economics Institute said the first place to look for capital is not at the bank or from some investor but from your own under-utilized capital.

He suggests an annual garage sale where everything that has not been used that year is auctioned off and turned into cash. "The return on capital can be improved by raising profits and by finding ways to run your business with less capital. Every dollar of fat carried around on your company's balance sheet makes it that much more difficult to achieve your objectives," Rutledge said.

Pasture Profits
The primary reason we aren't rich...
* Too much money sunk into land and machinery inhibits wealth.
* Investment should be concentrated in livestock.
* Work to maximize gross margin.
* Try to minimize non-livestock capital investment.

Chapter 9

The Care and Feeding of Your Customers

"The man who wants to do right will gain a rich reward. But the man who wants to get rich quickly will fail." Proverbs 28:20

P art of the crapshoot culture the stocker business developed is an unwholesome secretiveness in selling feeder cattle. This was probably a result of the buy 'em cheap, sell 'em high, buyer beware syndrome that glorified selling cattle for just a little bit more than they were worth. Some of these "marketing tricks" are not only unethical but downright illegal. Therefore the onus today is definitely for the grazier to step away from the crowd and "sign" his work as a grazier by putting his name on it publicly. It always amazes me that at board and video sales that are perfect to showcase a grazier's management skills to a wide audience, many offering cattle for sale refused to be identified by name.

Serving the Customer

Too often forgotten is that the cattle business is a people business first. There are no "reputation cattle." There are only cattle people with reputations. Everybody in the industry longs to do business with people they know and can trust. The best way to break out of the "commodity" market is to become a very real flesh and blood, caring human being to your customer.

I once read that the difference between a small business person and an entrepreneur is that a small business person goes into business for himself and the entrepreneur goes into business for others. If you can structure your operation so that its primary goal is to serve the customer's needs rather than just your own, you will virtually guarantee success. We all need other people to be able to stay in business today. If that need can be filled by someone in a pleasant, non-adversarial way, then that someone will get our business, support, encouragement, and many times, personal friendship. So what are our customers needs?

Large Feedyards

Today's large commercial feedyards need healthy cattle that can all be fed the same way with a minimum of individual care and attention. Commercial feedyards are feed manufacturing and marketing organizations. They buy raw feed ingredients, manufacture them, and then sell the feed to cattle owners who are rooming their cattle in the feed manufacturers' hotel. It is a very narrow-margin, high-volume business. Many commercial feedyards have less than one employee for each 10,000 head on feed, including the feed mill crew and office staff, so cattle disease "wrecks" and cattle needing individual feeds or feeding management have to be kept to a minimum.

While these feedyards will feed calves for others, they prefer to buy yearlings due to their better health and lower management requirements. If you can consistently provide yearling cattle that come in healthy and are already trained to eat at a feedbunk and drink from a waterer, you can quickly move to the front row of a feedyard's cattle suppliers.

Identifying Feedyard Needs

As far as genetic makeup of the cattle, it is the consistency of the makeup that is most important. The key need of the feedyard is that they be of an age and genetic consistency that they can be fed the same way. While the premium priced "Number Ones" get the most media attention, cattle feeders run value of gain analysis too, and frequently find the premium priced breeds offer the lowest value of gain. You can sell straight-bred Brahmans, Holsteins, Longhorns, or whatever as a result, as long as they are packaged in uniform weight, truckload lots. Some feedyards are tied in with packers and will only buy Choice grade cattle. Some only buy Angus crosses for the Certified Angus Beef Program. Part of your job as a grazier is to identify who feeds what, where and when and sell them what they want to buy.

For example, some Plains and Midwestern feedyards do not want Brahman-crosses in the Fall of the year, but do want them in the Spring. Some California and Arizona desert feeders want only Brahman-crosses. Some upper Midwest feeders only feed Holsteins, etc.

A big part of your job as a stocker-grazier is to assemble uniform-sized and weighted groups from the mish-mash mix of cattle available in your local area. Order buyers perform this function as well, but graziers have the advantage of owning the cattle long enough to virtually insure their good health upon arrival, a benefit most order buyers can't offer. This length-of-stay advantage has a real benefit with heifers too.

Selling Heifers

A pregnant heifer is a real management headache for a feedyard. As a result, many feedyards avoid the whole female sex. If you can guarantee all of your heifers are open, you can quickly get even the biggest feedyard's attention. This should not be a big deal for many graziers. If you graze a heifer for six months and own no bulls and have reasonably good fences, you can guarantee your heifers are open.

This is exactly what Moody Kennedy of Canton, Mississippi, does. He will send a feedyard $100 for any of his heifers they find bred. This is really putting your money where your mouth is. It has worked well for Moody whose yearling heifers are always in good demand.

Many states require that heifers from non-Class A Brucellosis states be spayed before they can be imported. If you are in a

Class C state, this may be a program you want to look at as your local heifers are probably carrying a discriminatory discount that you could turn to your advantage with a spaying program. Many graziers have found they can sell spayed heifers for virtually the same price as steers.

Graziers who use management-intensive grazing techniques have another advantage in that their cattle are already acclimated to tighter herd densities and feed considerably better than extensively managed cattle. Point this benefit out to the feedyard manager to further differentiate your cattle from the crowd.

Other Services

Another service you can offer is to feed the cattle a commercial receiving ration feed prior to shipping the cattle. With this program, the cattle are weighed off pasture, pencil shrunk, and ownership is transferred. Then the cattle are fed a locally purchased commercial stocker-feeder receiving feed at cost plus a markup or yardage fee for a customer-specified period of time, usually a few days to a week or so.

By selling the cattle prior to feeding them, any compensatory gain in the cattle accrues to the customer rather than the grazier and avoids the "warmed up" cattle controversy. This allows the cattle that have been grazed on lush forages to "dry out" and become feedlot adapted and accustomed to bunk feeding prior to being shipped. This service is particularly appreciated by feeders who are feeding cattle for their own account and can become another little profit center for the grazier.

Personal Contact

Contrary to popular notion, the building of a better mouse-trap does not insure that the world will come beating down your door. It still has to be sold. Remember our original goal was to become a real living, breathing person to our feedyard customer. The best way to do this is by personal contact.

Over half of the feedlot capacity in the United States is in a narrow band stretching northward from Lubbock, Texas, to south-western Nebraska. Skip over to Colorado and you can add another 15 percent.

It would be logistically possible in the course of a week's driving tour to visit feedyard managers with annual needs of millions of feeder cattle. But don't just drop in unannounced. You can have them expecting, even anticipating, to see you through a prior direct mail campaign.

Texas, Kansas, Nebraska, Colorado and a few other states have cattle feeding associations that will be glad to send you the names and addresses of their custom feedlot members.

Write a letter to all, some, or a few of them, explaining that you are a grazier looking for a custom feeder and are interested in knowing about their programs for cattle and feed financing, risk management, and joint venture and partnering programs. From those who respond choose a few, write or call for an appointment, and go see them.

Ask what kind of cattle they like to feed. Ask about problems they have with cattle from your area and what a person can do about them. Let them know you are concerned about their problems. Have a picture booklet of your place with you showing your grass, corral and working facilities, cowboys, cattle, horses, dogs and your spouse and kids. Most good cattle people, feedyard or grass, want to see another person's ranch and grass conditions. Tell him about any unique services or guarantees you may have. He'll also ask about your animal health program, what shots you give, and so on.

Partnering Program

What you are trying to do is establish a bond that you are a real person with a spouse and kids who is trying to make a little money in the cattle business just like he is. If it seems like you both hit it off well together, tell him you would like to partner with him on a truckload, or a pen of cattle.

In a partnering program, the feeder buys a percentage of the cattle from you as feeders. You retain ownership to a certain percent, but they are fed together, so if they are a good set of cattle both parties benefit.

Feedlot partnerships do not necessarily have to be a 50/50 split. They can be a 95/5 split where you only retain ownership on five percent, a 90/10, or whatever. There is little advantage to a grazier in retained ownership unless feeder cattle are discount to fat cattle and the feedyard manager understands this. All you are trying to indicate is your goodwill, faith in the cattle, and a willingness to put your money where your mouth is.

Pricing Feeder Cattle

The pricing of feeder cattle is at best a difficult science and can turn the best relationship contentious. All any feeder can do is offer the going price for feeder cattle in HIS area. You may be able to get more for your cattle in another area, but in most cases you will find the premium is exactly equal to the increase in the freight to get them to that market, so be careful about chasing premium prices.

Some graziers will send a group of cattle through their local auction sale to establish a base price for negotiation. Others use Cattle-Fax, DTN and other cattle price quote services as a baseline. A consistently fair price from someone you know and trust is usually the best price.

The above may sound like a lot of trouble for such a small reward, but being a somebody to a man who buys feeder cattle in the hundreds of thousands can have a lot of fringe benefits. There are hundreds of millions of dollars looking for competent grass and cattle management. If you can become known as one of those people who can consistently produce good cattle performance on grass, you can get all the cattle and money you can handle.

Pasture Profits

* The cattle business is a people business.
* Sign your work.
* Become a real person to your customers.
* Try to do business with those who do business with you.

Chapter 10

Cows and Stockers Together

"An empty stable stays clean, but there is no income from an empty stable." Proverbs 14:4

Beef cattle are being raised around the world for a small fraction of the North American cost of production. True grass farmers recognize that while grazing animals can get painfully cheap, the grass has absolutely no value without them.

In North America, it is difficult to see the true economic role of the beef cow because production has been largely subsidized by other sources of income. However, if we look at the non-subsidized production areas of the world, we find that in temperate grass climates, the beef brood cow is primarily used for breaking rough land and to remove overly mature forage from ewe/lamb, dairy and finishing class (stocker) beef pastures. It seldom is the centerpiece operation.

On a worldwide basis over 70 percent of the world's beef cows are located in tropical and subtropical climates. These climates offer a year-round supply of forage, but much of it of low quality. These cows' male calves are traditionally weaned and shipped to temperate grass areas where they can be more quickly grown and finished on grass. In Argentina and Australia it is common for ranchers to have two ranches — one for the cows in a hot, wet climate and one for growing and finishing in a cooler climate capable of growing temperate grass.

Opportunity Cost

On very high quality pastures, the biggest cost in cow-calf production is the opportunity cost of not running much higher-return sheep, dairy or stocker cattle there. This does not mean that a stocker operator should not have a cow herd, nor that a cow-calf producer shouldn't vertically integrate and/or buy stocker calves. It does mean that the cow should be relegated to a value-adding role. The grass rule is "The less you value your forage the more you will value a beef cow."

For example, the semi-arid climate of the Southwest can produce an excellent year-round cow-calf type of forage. The dry climate produces a nutrient-rich grass that cures well as standing hay for winter dry-cow feed. Unfortunately due to the wide variation in year-to-year rainfall, the use of cows alone to harvest the forage results in a tremendous forage waste if stocked for the bad years, and high feed costs, long-term range deterioration and forced destocking if stocked for the good years. Many of these producers have found that the addition of a seasonal spring-to-late-summer stocker program allows them a much more flexible way to harvest their grass than if they had cows alone. These stocker programs are timed to the naturally-occurring high-quality spring grass periods and require no additional expense or manipulation to produce. However, these ranchers are still primarily cow-calf producers in recognition that their forage resource lends itself well to beef cow-calf production. During forage shortages, it will be the stocker operation that should be discontinued.

In contrast, a producer in a short-season, temperate grass region may produce enough by-product hay and silage just in the act of maintaining young, stocker-quality pasture to winter a small cow herd. Beef cows add value to this otherwise wasted forage resource, but he is still primarily a stocker producer. If there is a drought and no surplus grass is available for hay, it is the beef cow operation that will be liquidated.

In long-growing-season areas, if most of us will look at our grass production as a season-long whole, we will see that a combination of classes and species usually optimizes our grass harvest. However, the enterprise that best fits the bulk of our grass quality and production should predominate and the other enterprises should complement this primary enterprise. What has been overlooked is that once we decide to have a post-weaning grazing program we can dramatically cut the costs of the beef cow.

Early Weaning Increases Efficiency

Auburn University Extension Nutritionist, B.G. Ruffin, pointed out that it takes two and one-half times the TDN to produce a pound of beef before weaning as it does after weaning. Total forage harvest efficiency is not increased by keeping the calf on the teat longer, but rather by the reverse.

The producing of heavy weaned calves in the fall is not grass efficient. Lighter calves have a lower body maintenance requirement and can be overwintered cheaper. Therefore, a light weaning weight becomes a positive rather than a negative. This is best accomplished by calving later in the year and weaning earlier. Calving later in the year has other important benefits as well.

Late Calving is the Key to Cutting Cowherd Costs

Several years ago, consulting nutritionist Dick Diven was hired by a feedlot client to come up with a nutrition program to make the feedlot's ranch customers more profitable. Dick said the first thing that hit him was that an 80% calving rate was considered to be "good." He didn't see how any ranch could be made economically viable with 20% of its cow herd not producing a calf. Therefore, he began to investigate all the factors that impacted on a beef cow's reproduction. Two major factors Dick discovered were day length (latitude effect) and body condition score at calving.

While it had been thought that the seasonality effect so prominent in sheep had been bred out of beef cows, Dick found this was not entirely true. Annual cow sexual activity was found to be the highest at the spring and fall equinoxes (mid-March and mid-September). However, the shortest postpartum recovery period was found to be in July, or immediately following the longest day length of the year.

For example, the postpartum length at 40 degrees latitude (northern Missouri) was found to be only 31 days in July for a cow in Body Condition Score of 6.5 versus a postpartum length of 69 days in February.

A short postpartum recovery period is very critical in maintaining a 365-day calving cycle. Dick said that a breeding season longer than 45 days would always tend to creep later and later each year. This is because cows that calve during the latter part of a 63-day or longer breeding season are not capable of conceiving any earlier in subsequent seasons.

Heifers born in late spring or early summer mature sexually 2.5 to 3 months earlier than those born during the winter months. For a heifer to calve on her second birthday, breeding must occur by 15 months of age. The heifer born in May, June or July did not have to breed until at least their third post pubertal estrus, whereas, heifers born in December must breed at pubertal estrus.

Research in Kentucky found that delaying breeding in heifers for two normal cycles could increase the conception rate by as much as 21%!

Dick figures that a 96% conception rate could be an attainable goal for summer calving heifers.

In addition to day length, the cow's reproductive rate was found to be very closely tied to her level of body fat at calving. A slightly fat cow at calving had a much higher chance of quickly breeding back than a moderately fleshed cow. The pregnancy rate of a cow in a moderate body condition score of 5 was nearly half that of a cow in a nearing obese body condition score of 7.5. (One body condition score is equal to about 90 lbs of body weight.)

Dick prefers easy fattening breeds of cattle such as Angus and Hereford for least-cost cow-calf production. He said the annual gain and loss in body condition in a beef cow, which many see as a "waste," is actually a major cost reducer if it is used to replace purchased feed. Body fat is the most readily available energy source for the cow, and it is energy that is the critical factor in rebreeding a lactating cow.

Unfortunately, body fat is very expensive to produce. It requires 2.95 times as much energy to produce one pound of fat as that one pound of fat can give up as energy to the cow. However, it is very difficult for a lactating cow to gain body condition, and this is particularly true with heifers.

The only cheap time to put on bodyfat is during the pasture lush, and the most reliable way of insuring a fat cow or heifer prior to calving is to delay calving for two breeding cycles past the normal onset of spring pasture. In most areas of the country, this will be the period of June to August. This also corresponds to the shortest postpartum recovery period.

The one exception to this rule is the desert Southwest where the pasture lush occurs following the summer monsoon season. In this area of the country, Dick recommends a fall calving season starting in September.

While many beef cow graziers try to time their calving just prior to the onset of the spring lush like dairy graziers do, this sets them up for major problems in years with dry or unusually cold springs. Dick said it is much safer to time the breeding season later in the spring when grass was assured than earlier when it was iffy.

He said the first 13 days are the most critical period in a calf's life nutritionally. A calf denied adequate milk immediately after calving will never develop normally and will be a dogie, or a "short," its whole life.

Another critical nutrition period is when a growing beef animal is at about 65 percent of its bodyweight. This normally occurs at 12 to 13 months of age. The animal must be gaining weight at that time for it to have the ability to deposit intramuscular fat and grade Choice at slaughter. The later calving also ensures the pastured yearling will be gaining well during this critical stage as well.

Now, let's sum this up.

Arguments For Early Summer Calving

☆ Lowest requirement for supplemental feed.

☆ Shortest postpartum recovery period.

☆ Heifers mature two to three months earlier.

☆ Both cows and heifers easier to keep on 365 day cycle.

☆ Cows fatten cheaply on spring lush.

☆ Yearlings assured of good gains during "critical" period.

☆ Increase in calving percentage by 15 to 20% over winter calving.

Arguments Against Early Summer Calving

☆ Lighter weaning weight in the fall due to younger calf.

As we previously discussed, this one drawback is not a drawback to a stocker grazier. However, Dick said you don't necessarily have to wean the calf in the fall. As little as one pound of milk can provide all the protein supplement a 400 lb calf needs on frosted range, albeit, the calf will lose weight.

Thanks to the "miracle of compensatory gain" as Gordon Hazard calls it, supplementing calves to gain weight in the winter is of no economic value if they are to be retained through yearling age. Dick pointed out that calves weaned at five months of age will weigh the same at the same age as calves weaned at seven months of age once they reach 65% of mature body weight or around 12 to 13 months of age.

This weight convergence occurs because the heavier calves will use more energy for body maintenance while the lighter calves will devote more energy toward gain. Therefore, there is no advantage to a heavier weaning weight as long as we retain ownership through the yearling phase.

The post weaning weight gain is the easiest to put on because it is mostly protein and water and requires very little energy. As a result, the stocker phase is the most profitable phase of the calf's life.

Dick said it is wrong for vertically integrated ranches to economically measure the cow-calf portion as a separate enterprise from the post-weaning stocker portion as this constantly reinforces the old, high-weaning-weight, high-cost paradigm. For a least cost approach, the cow-calf and stocker phases should be seen and managed as one seamless enterprise.

Pasture Profits
* It is the quality of the forage that separates stockers from cow-calf production.
* Running beef cows on stocker quality forage has a very high opportunity cost.
* Either cow-calf or stocker should be the dominant enterprise and all other enterprises should compliment the dominant enterprise.

Chapter 11

Pasture Popping — Maximizing Compensatory Gain

"A wise man thinks ahead; a fool doesn't and even brags about it." Proverbs 13:16

Pushing stocker cattle to a higher rate of gain with supplemental feed is not cost effective if the cattle are going to
be owned through a subsequent grass period when high
daily gains would come naturally. Growing cattle have the amazing
ability to "catch up" or compensate for slower growth rates with a
subsequent higher rate of gain when forage quality increases. This
is usually called compensatory gain or "pop."

Yearlings will have a lot of this unexpressed growth in them
if they have grazed warm-season tame pastures, full-season range,
endophyte-infected fescue, or have been wintered on short small
grain, or winter pasture. Gains of four to six pounds a day are not
uncommon when cattle of this type are swapped to higher quality
forages.

Research at the Kerr Center in southeastern Oklahoma has
shown gains of six and one-half pounds a day when yearlings that
had been wintered on endophyte-infected fescue were swapped to

annual ryegrass in the late spring. Planning to "finish" your stocker cattle on a high-quality "poppin'" forage prior to sale is good stocker management. But it requires forage planning.

Growing cattle are most profitable when managed to grow heavier on an ever-rising plane of forage quality. Such an inclined plane maximizes the use of compensatory gain.

Correct Forage Sequence Important

For example, a forage sequence capable of growing a calf from 200 lbs to 1000 lbs might be as follows: warm-season perennial, such as bermuda and bahia; followed by cool-season perennial, such as fescue and perennial ryegrass; followed by warm-season annuals, such as crabgrass, dwarf millet or grain sorghum; brassicas, such as rape and kale; or legumes, such as lespedeza, kudzu, cowpeas, fine stemmed soybeans, or alfalfa; followed by a cool-season annual, such as oats or wheat.

This sequence should be planned so as to overwinter the animal only once in your ownership. Do this early in the animal's life, at as light a weight as possible. In the above example, the animal would be "finished" or "popped" on the oats in the fall, but not overwintered on them.

Forage Quality

Cool-Season Annuals
Warm-Season Annuals
Cool-Season Perennials
Warm-Season Perennials

The key to remember is the forage sequence. Warm-season perennial, cool-season perennial, warm-season annuals, cool-season annuals. A reverse sequence, such as cool-season annuals to warm-season perennial can actually cause the animals to standstill or lose weight because you have lowered the digestibility plane. Costly annuals are most cost-effective when used with yearling age cattle.

Cowpeas and millet, for example, produce no better gains on calves than a far cheaper warm-season perennial like bermudagrass. Most warm-season perennials are of "finish" quality until the weather turns extremely hot, after which they become "framing forages" to ready cattle for a subsequent higher quality cool-season graze.

Gain Per Acre Versus Gain Per Head

Remember, gain per acre and gain per head are the opposite sides of a teeter-totter. If one goes up, the other comes down. Allowing cattle to become too fleshy at an intermediate stage not only hurts the subsequent "finishing" performance but indicates you have given up too much in gain per acre. We only want to fully express our cattle once, and that is on the last forage sequence. This allows us to maximize gain per acre with the intermediate forages and still produce a high overall average daily gain.

Research at LSU showed that the grass gains between 700 lbs and 900 lbs were more profitable than those between 400 lbs and 700 lbs because there was much less price rollback. This produced a higher value of gain.

Once you reach 700 lbs with an animal, you've already absorbed the worst of the price slide. If you've got the forage available, you might as well keep going.

Popping Paddocks Pay

Setting aside a couple of paddocks in the early spring and allowing them to accumulate growth for "finishing" or "popping out" stocker cattle you are preparing to sell can pay big dividends according to R.L. Dalrymple of the Noble Foundation in Ardmore, Oklahoma. In the spring of 1991, Dalrymple allowed 77 head of wintergrazed stocker cattle to express their built-up compensatory gain on two purposefully deferred paddocks that had built up a lush stand of rye/ryegrass.

These cattle gained an extra 33 pounds per head in just six days for an adg of 5.5 lbs. This was an increase of 106 lbs of beef per acre or 2541 lbs for the 24 acres in the two deferred paddocks.

Dalrymple said two 12 acre paddocks of rye/ryegrass were

purposely deferred in the early spring and allowed to grow up to 12 to 18 inches in height. They were rotated off after two days when a three- to six-inch residual still remained. Another herd of stocker cattle on non-deferred shorter pasture gained only 1.8 lbs per day in the same time period, or 831 lbs versus the 2541 lbs produced on the deferred pasture. "The deferred finishing paddocks produced 710 lbs more beef than the non-deferred paddocks," Dalrymple said.

"At a value of gain of 60 cents a pound that's an increase in net profit of $42.72 per acre or approximately eight times the per acre cost of subdivision," he said. "It is these kinds of things we can do with pasture subdivision that we just can't do with continuous grazing."

Pasture Profits
* Stocker cattle should always be moved to a forage higher in quality than the one they left.
* Compensatory gain should be maximized.
* Forages must be used in the correct sequence to maximize gain.

Chapter 12

What's the Correct Stocking Rate?
It Depends...

"The shrewd man does everything with prudence." Proverbs 13:16

The Bible reminds us that it is the eye of the master that fattens the cattle. A good "grass eye" is one that can match animal nutritional needs with the grass growth and forward supply. This matching of animal and feed supply is commonly expressed as the stocking rate.

At the International Grasslands Congress is Saskatoon, Saskatchewan in 1997, the most common theme was stocking rate, stocking rate, stocking rate. Speaker after speaker emphasized that stocking rate is the over-riding factor of profitability. On that one point there was unanimous agreement. The disagreement came about over how to create this "ideal" stocking rate.

Many speakers blasted rotational grazing as worthless. Others vociferously defended it. Recent research into stocking rate at the University of Missouri Forage Systems Research Center might illustrate why these seemingly opposing viewpoints could both be correct.

"We have found no effect good or bad on animal perfor-

mance from rotational grazing early in the season," said Jim
Gerrish, the coiner of the term "management-intensive grazing."
"For the first 70 to 90 days there is no difference between grazing
systems. The difference is completely between stocking rates."

Gerrish said given enough grass, a low to intermediate
stocking rate, steers that have been over-wintered on hay have
gained between 4 and 5 pounds per head per day for the first 42 day
weighing period on 30 percent infected fescue. "The crunch comes
out there in the summer when the grass growth slows down. It is
not until that point that rotational grazing becomes superior in
terms of animal gains."

In other words, rotational grazing has no benefit on animal
performance as long as the grass availability is in excess of animal
demand.

Rotational grazing is a rationing tool. All it does is ration
out a stockpiled crop or restrict animals so that a stockpile can be
built up in front of the animal. This is why New Zealand extension
spent 20 years teaching autumn stockpiling and winter rationing
before they touched the summer growing season.

In contrast, in North America we have been trying to start
novice graziers off with the green growing season when rotational
grazing's usefulness is not nearly as clear-cut and the degree of
expertise required for success is extremely high. It is true that a
highly skilled grazier can use management-intensive grazing methods to move surpluses around during the year, but this is a skill
level we might hope to obtain in a lifetime, not in an afternoon's
training.

Note that both of the above valid uses of rotational grazing,
restricting consumption to build a forward stockpile of grass and
rationing out stockpiled grass, lowers animal performance in the
short-term. Therefore, if maximum gain per head is your sole
management goal, you don't need to be doing rotational grazing.

For example, many graziers were dismayed when National
Farms tore out the Devore Ranch grazing cell in the Kansas Flint
Hills after they bought it. For the 60 day late spring/early summer,
high stocking rate, "popping-out" type of grazing they wanted to

use the ranch for, set stocking was probably the correct grazing method. However, they still should have left the paddock subdivisions, as these allow a much more accurate matching of animals to pasture growth and availability, even with set stocking and would allow full-season use of the ranch if that ever became a management goal.

However, deciding what method of grazing management we are going to use does not answer the question of "What is the correct stocking rate?" Not only is the grass growth rate highly variable within the season, but between seasons and from one year to the next. In much of North America, there is a 300 percent variance in annual forage production due to rainfall. Also with growing animals, the animal is lighter at the start of the season than he is at the end of the season or just backward of the way the grass growth curve behaves. Considering these variables, the only correct answer is "It depends."

It depends upon the weather, the season, the amount of market rollback per head, and whether or not you are going to own the animal through another phase or season. In other words, the goal of grazing management is not so much to determine one stocking rate for the whole season, but to monitor animal and grass growth so as to create a variable stocking rate.

In arid and semi-arid climates, the variation within the season is not as great as it is between years. Therefore in these climates stocking rate must be based upon animal removal. This is why dry country beef ranches have traditionally been structured as cow-calf/yearling operations.

In a good year, the calf crop is retained and grown as heavy as possible to utilize the grass and in a bad year they are sold as light as possible to reduce the stocking rate. In some areas of the Southwest where multiple-year droughts are common, a complete and total destocking may be necessary. What is both economically and environmentally destructive is to leave the animals on the land and buy in feed.

Several African speakers at the International Grasslands Congress said that prior to such intervention efforts there was never any lasting damage to the range because the animals died before the plants. Therefore, it was their view that there can be no lasting damage from overgrazing without supplemental feeding.

Why It's Called Management-Intensive Grazing

In contrast, in humid climates rainfall averages from year to year tend to be relatively stable but there is a tremendous amount of difference in forage production within the growing season. This is doubly true at the higher latitudes of North America.

Such extreme within-the-season variation makes determining the correct stocking rate much more difficult and requires a very high level of monitoring and quick reaction. A stocking rate that is correct for June will be way too heavy in August. Therefore in humid climates, grazing management must concentrate on methods of creating highly variable within-season stocking rates more than a single season-long one.

There are four main ways this variation in stocking rate is accomplished:

1. **Buffer Grazing.** This is the most common system used in Europe for growing beef animals. With buffer grazing the animals are set-stocked and a portable electric fence is used to create a variable stocking rate by giving an increasing portion of the paddock as the grazing season progresses. Buffer grazing is the preferred method of grazing management for very young animals (up to 350 lbs) and is an excellent tool for small acreage stocker graziers. Its drawback is that it does not allow for building up a stockpile for wintergrazing and should only be used with sod-forming grasses like bluegrass, perennial ryegrass and bermuda.

2. **Rotational Grazing.** With rotational grazing, the land area is subdivided into paddocks. An increase in stocking rate is created in the spring by dropping paddocks — as much as half of the land area — from grazing and mechanically harvesting the surplus for use during forage short times. Most of this is used as winter feed but can also be used to fill shortfalls in the summer and fall as well.

Rotational grazing is the most common variable stocking

rate tool where a relatively constant number of animals must be maintained through the whole year. A good example of this is a dairy. Dairies are traditionally located in areas with reliable rainfall or utilize irrigation due to their relatively inflexible stock numbers. In New Zealand, dry stock and young stock are traditionally grazed off the farm to allow the maximum conversion of grass to milk.

In the Northern Hemisphere where dairy cows are commonly supplemented with grain, decreasing or eliminating grain feeding to dairy cows during surplus grass periods also helps to create a variable stocking rate effect. This is a lesson yet to be learned by most North American dairy producers.

Rotational grazing allows the use of more upright growing, hay-type grasses and alfalfa. These plants can produce more dry matter per acre and are more drought tolerant than the shallow-rooted cool-season sod grasses.

Two major problems with rotational grazing are its reliance on costly machine harvest and the necessity for a large percentage of the farm to be machinery-accessible. Due to the harsh effect of total plant removal (via hay or silage) on the grass and soil, acreages being dropped for machine harvest needs to vary every year.

Most beef farms and ranches tend to be located on machinery-inaccessible land and so this system needs to be modified to include a variable number of grazing animals.

3. **Heavy Scavenger Steers**. Under this system, heavy steers are purchased to utilize the 50 to 70 day grass surplus in the spring. By buying feeder weight (750 lbs plus) animals in May, price margin rollback is minimized as these are normally the lowest priced cattle of the year due to seasonally high grain prices. Negative margin risk is much higher with heavy cattle and these "scavenger steers" should always be forward sold (booked) at the time they are bought.

4. **June Calving Scavenger Cow Herd**. Because a beef cow's forage intake dramatically increases at calving, timing the onset of calving to the seasonal grass surplus can create a quick increase in stocking rate. These cows and their calves can then be used to harvest dropped, machinery-inaccessible paddocks that

have gotten too mature for good steer gains.

By calving late, these cows will have enough fat on their back to keep the calves' gains high while on the poor quality forage. These cows' calves are weaned as soon as the surplus is gone and the cows are dropped to maintenance. The secret of this program is to not calve before you have a genuine grass surplus.

Remember, once the decision is made to forward integrate through the stocker phase, the post-weaning portion of the animal's life must become the centerpiece activity. A high weaning weight should not be a goal. The stocker portion is so profitable because most of the gain is water and requires very little energy. The beef cow is always the most profitable when she is used as a value-adding scavenger of lower quality forage resources as in the above examples.

The creation of a system of variable stocking rates still does not answer all the questions necessary to determine the stocking rate. For example do you want a high gain per head or per acre?

What Rate of Gain? When?

Gain per head and gain per acre are always at the opposite ends of the teeter-totter. If one goes up, one must go down. Here are a few guidelines to help you choose which rate of gain to use and when.

Use a High Gain per Head When:
1. You have a large negative price margin to absorb.
2. The animal is nearing final sale or slaughter.
3. You have more land than capital for cattle.
4. Corn is expensive and feedlot time needs to be minimized.

Use a High Gain per Acre When:
1. Price margins are narrow or positive.
2. Animals will be owned through another phase or through the coming winter.
3. You are selling gain by the pound.
4. Corn is cheap enough to grow as well as finish in the feedlot.

With growing beef cattle it is more cost-effective to use compensatory gain than to try and keep an animal at a high rate of gain for his whole life. In theory, you would only want to "pop" an animal once in his life to keep body maintenance requirements low and gain per acre high. In grass finished beef countries this "popping" is timed to occur just before slaughter. However, at the Forage Systems Research Center, steers that were allowed unrestricted access in the spring kept their gain advantage over restricted steers for the whole grazing season.

Unfortunately, the low stocking rates, which produce high animal performance, tends to also produce low-quality, poorly tillered, open grass swards. Jim Gerrish told me that at only a half steer per acre, his paddocks quickly lose their legume component. A legume component is particularly valuable to keep the energy content of the forage high during the "summer slump" when the grasses lignify and decline in digestibility.

The desire to have both high animal performance and a high level of production per acre has led many to adopt leader-follower grazing systems.

Leader-Follower Systems Increase Management Intensity

For example, in Argentina steers nearing finish are sorted from the "growing herd" and allowed to graze in front of the main herd in a leader-follower method. This system was formalized in the 1930s as the "Hohenheim Grazing Method" and often included three classes of growing animals — calves, yearlings and a finishing class as the front grazer. Needless to say this was an incredibly management-intensive management system and today most graziers only use two classes.

If lactating cows are used as a follower forage cleanup herd it is critical that they not calve until there is a surplus of grass or they will be hard-pressed to rebreed from the low energy residual left by a herd of steers. With breeding animals peak nutrition needs always must be timed to the peak of the grass cycle. The primary management goal with beef cows is to get the cow bred. We don't need a cow to grow the calf, but only a cow can create one.

I have included below a "what if" stocking rate scenario provided by Jim Gerrish and the Forage Systems Research Center that can help you grasp the effect of stocking rate on return per head versus return per acre. Note that when stocking rate went over 1.5 steers per acre the return per acre and the return per steer fell. As Peter Drucker has noted, "The optimum is always 75 to 80 percent of the maximum."

Impact of Stocking Rate
On Gains per Acre, Gains per Head, and Profits

Stocking rate (hd/A)	0.5	1.0	1.5	2.0
ADG (lb/hd/day)	2.4	2.0	1.6	1.2
Gain/acre (lb/A)	252	420	504	504
Purchase weight (lb/hd)	400	400	·400	400
Purchase price/cwt ($/cwt)	$83.60	$83.60	$83.60	$83.60
Purchase cost/head ($/hd)	$334.40	$334.40 .	$334.40	$334.40
Variable costs/head ($/hd)	$30.00	$30.00	$30.00	$30.00
Operating interest ($/hd)	$20.38	$20.38	$20.38	$20.38
Total cost/head ($/hd)	$384.78	$384.78	$384.78	$384.78
Sale weight (lb/hd)	904	820	736	652
Sale price/cwt ($/cwt)	$64.75	$66.75	$68.75	$70.75
Gross revenue/head ($/hd)	$585.34	$547.35	$506.00	$461.29
Net revenue/head ($/hd)	$200.56	$162.57	$121.22	$76.51
Net revenue/acre ($/hd)	$100.28	$162.57	$181.83	$153.02
Fence & water cost ($/A)	$39.82	$39.82	$39.82	$39.82
Fixed costs ($/A)	$11.73	$11.73	$11.73	$11.73
Profit/acre ($/A)	$48.73	$111.02	$130.28	$101.47
Value of gain ($/lb)	$0.498	$0.507	$0.511	$0.504

Profit (return to land, labor and management as used here) projections for various steer backgrounding scenarios.

Source: The Forage Systems Research Center

No doubt I have only scratched the surface of the list of variables that should be considered in determining stocking rate and grazing method. For example, Gerrish said in his heavy clay soils he would not use rotational grazing in years with wet springs due to high soil compaction and pugging damage, but he would in dry spring years. In "normal" years, he said he might start out with continuous grazing and not start rotating until mid-summer. "(With beef cattle) there are valid times to rotate and equally valid times to not rotate," he said. "There are reasons to have a high rate of gain and equally valid reasons to have a low rate of gain. Grazing is a game of managing a highly variable whole. That's why I said we ought to call what we do `management-intensive grazing'."

Pasture Profits
* Stocking rate is the over-riding factor of profitability.
* Rotational grazing becomes most beneficial when grass supplies are greatly diminshed in relation to animals' needs.
* The goal of grazing management is to monitor animal and grass growth in order to create a variable stocking rate.

Chapter 13

Heifers, Bulls, And Dairy Calves

"He who loves wisdom loves his own best interest and will be a success." Proverbs 18:8

G rowing animals with forage does not have to be limited to beef breeds. During the high price phase of the cattle cycle, it is often more profitable to switch to grazing females as replacements either for beef or for dairy or to buy lower priced dairy breed males. What you should graze and when will be determined by your value of gain analysis.

Grazing Management for Beef Replacement Heifers

Stan Parsons, of Ranch Management Consultants, said that a rancher can always buy his replacements cheaper than he can grow them out himself. This is undoubtedly true for areas of the country where winter gain is expensive to put on. But in areas of the country with good winter pasture, a replacement heifer development program can provide a good shelter from the negative margin storm at the peak of the cattle cycle.

Remember, at the peak, fat cattle fall first, then feeders, then calves. This means that the average cow-calf producer does not really feel the market drop until some three years after the peak price in fat cattle. Traditionally, the greatest percentage of the herd building occurs **after** the peak of the cattle cycle. At this time bred heifers may sell for several hundred dollars more than feeder steers, which are already on their way down in price.

However, this is a relatively short-term event.

Always keep in mind the axiom, "No heifer ever conceived during a period of high cattle prices ever had a calf that was sold during a period of high cattle prices." This axiom recognizes two facts: one, the peak price cycle is relatively short; and two, the replacement cycle is relatively long. The period from conception to a salable calf often is as long as 50 months. The high price cycle is typically only 36 months in length.

A stable to rising market in replacement heifers in the face of a falling fat and feeder steer market means it is time to not replace your bred heifers with more heifers but with steer calves as the party is soon to be over for cow-calf producers — too soon for you to grow out and breed another set of heifers.

To cash in on the replacement heifer market requires a shift to heifer calves **before** the demand for replacements is evident. This time to shift will be indicated by a growing negative margin between yearling steers and their calf replacements. Remember, let the margin be your guide.

For you as a grazier it is important to shorten the period between weaning and conception to keep cash flow turning. Most graziers do not want their heifer program to take much longer than their normal steer grow out time. This means breeding the heifer as a yearling rather than as a two-year old.

Many graziers have expressed disappointment with this "yearling" mating program due to both low conception rates on the first mating and even lower conception rates when trying to breed back a two-year old nursing a calf. New Zealand research has shown that virgin replacement heifers that were grown out at high rates of gain over the winter (high grass allowances) on ryegrass pastures attained puberty earlier, and were significantly heavier, both before and after mating, than heifers that were wintered at a much more moderate rate. However, the subsequent pregnancy rate in both sets of heifers were virtually the same.

Most heifers will show first oestrus at around 60 to 65 percent of their grown liveweight. Heifers must reach this critical minimum in order to breed, however, the New Zealand research indicated that there was no benefit in the cow's lifetime productivity by **exceeding** this minimum weight at first mating. English and English-cross cattle should have no problem arriving at this weight in the spring without having to push them on cool-season pastures.

Key to Breeding Percentage

The trick to a high breeding percentage on yearling replacement heifers is to put no restrictions on the amount of green, spring grass they can cram into themselves from the onset of spring until the end of the breeding season.

With an average liveweight of 550 lbs at the start of the spring lush, English yearling heifers were able to achieve an 85 percent breeding rate in a 42-day breeding period by set stocking on lush, six-inch-deep spring pasture. European breeds were able to do just as well if they arrived at the start of the spring lush at a weight of 625 lbs to 650 lbs.

While most graziers like to sell the heifers once they can guarantee them as bred, it is very important that you tell your customers what to do with these bred yearling heifers to prevent subsequent calving problems. Giving these young heifers a high pasture allowance in late pregnancy dramatically increases calving problems and calf mortality. These young cows should be kept reigned in and only allowed to gain around a half pound a day until they calve. Then, after calving, like the yearling heifers, they should be turned loose without restrictions on the amount of green grass they can eat until after the breeding season is over.

Short-Wean First Calf Heifers

New Zealand extension recommends that calves on two-year-old cows should be weaned when the calves reach around 180 lbs. This takes the pressure off the young cow and allows you to turn her in with the main cow herd, managing her for the rest of her life like a mature cow.

Breeding in Sync with the Grass is the Key to Cow-calf Success

New Zealanders have found the whole trick to successful beef cow-calf management is this timing of the calving season to the onset of the spring lush. They have not found any other management practice that even remotely approaches the importance of this in cow-calf profitability.

Some graziers worry about calving in the spring lush, because they think their cows will produce too much milk. But the New Zealanders have found that true beef breeds are relatively insensitive to grass quantity and quality in the amount of milk they produce.

Beef cows gain weight rather than increase milk production when grazed laxly. Conversely, beef cows milk about the same when pasture conditions are poor by losing weight.

The grass management program for beef cow-calf is simple, and practically fool-proof once the cow has been bred. In fact, pasture management for the beef cow after the breeding season has ended has been shown to have very little effect on the weaning weights of the calves. This is due to the cows' ability to balance growth with variable milk production.

New Zealanders wean beef calves at four to six months of age. The calf is a fully functional ruminant at that age. By weaning early, they are able to drop the cow to maintenance levels and save grass.

The New Zealanders have also found that dry beef cows can be managed at very low levels of nutrition (losing as much as 10 to 15 percent of their body weight) until two months prior to calving with no subsequent ill effects on the reproductive rate.

Bull Grazing Tips

Grazing bulls, whether for seedstock production or slaughter, can be a different kettle of fish than grazing steers. In New Zealand, Holstein bulls are their primary stocker animal. I asked Keith Milligan, a regional grazing manager for the New Zealand extension service what his recommendations were on successfully grazing bulls, and in particular the size herd or mob to use. I had seen everything from 20 head mobs to 200 head mobs in my travels down there. The following was Keith's advice.

1. **Soil Conditions.** The wetter the soil the larger the herd. This limits the total area of pasture damage to a small rather than a large area. Milligan said it was his experience that you are better off to have a large mob and shift less frequently than vice versa.

2. **Age of Bulls.** There are very few social problems with bulls until they are 14 to 15 months of age or older. Younger bulls can be grazed just like steers.

3. **Liveweight Gain Targets.** The higher the liveweight gain expected the smaller the mob we opt for. This reduces competition between animals and allows, especially in older bulls, the establishment of territory. Territory establishment is not as important in a rotational grazing system, particulary when animals are shifted on a daily basis.

4. **Pasture on Offer.** If pasture on offer is low and residual dry matter post-grazing is also low, go for larger mob sizes. Conversely, if pre-grazing pasture levels are high and residuals are high (and by inference liveweight gain is also high) smaller mobs are more successful.

5. **Total Pasture Mass.** When you are trying to build a forward feed bank (stockpile) for winter grazing putting the bulls in a large mob will help because a lower proportion of the total area available is being grazed at any one time. Conversely, in the spring when pasture mass is outgrowing the ability of the bulls to eat it, putting a small mob of bulls in each paddock will increase food intake per head and therefore reduce the risk of the quality of the pasture falling.

Slaughter Bull Grazing in New Zealand

Milligan said the Kiwi bull beef farmer traditionally buys 10-week-old calves in the fall of the year. He said with calves this young it doesn't really matter what herd size you use because liveweight gains will be low regardless of your management. Mob sizes of around 100 are used with these baby bulls and a liveweight gain target of about a pound a day is set until spring.

As the pasture growth rate picks up in the spring, the pastures are top-grazed to maintain quality and a fast rotation using very large herds of bulls is used. As the growth continues to increase the herds are split and resplit into smaller groups to maintain pasture quality and maximize live weight gain.

As growth slows in the summer and pasture quality falls, the difficulty in bull grazing increases tremendously.

Summer is Bull Fighting Season

Milligan said one of the problems with splitting a herd of bulls in the spring is that if they are then recombined, there is a mix of hierarchial social structures, which generally lead to renewed social problems. It is during the summer that major social problems with bulls start to occur. Most New Zealand graziers overcome most of these social problems by slaughtering bulls at 15 to 18 months of age. He said bulls over 15 to 20 months of age should be kept in mob sizes of 30 to 50 head. Once the weather begins to cool and the bulls start moving toward winter most of your bull fighting problems will end, and recombining the herds will be possible with few problems.

If grown bulls must be combined into a herd, many New Zealand graziers have found that spraying them all with a flyspray or an astringent and water mix will prevent fighting. Apparently if all the bulls smell alike fighting is reduced. In Europe, billy goats are kept with purebred bulls to both prevent fighting and riding and to serve as a "miner's canary" to diseases. Goats are more susceptible to most cattle diseases than the cattle themselves and serve as a warning of an impending disease outbreak by getting sick before the bulls.

Bulls, New Zealand Style — A Case in Point

On a river plain near Palmerston North on the North Island of New Zealand is what is no doubt one of the most intensive grassfed beef production systems in the world. The Burleigh Bull Beef System, devised by local grazier, Harry Wier, produces 1200 to 1500 lbs of beef per acre per year from intensively subdivided and irrigated pastures of perennial ryegrass and white clover.

The Wiers' 400 acre grass farm is currently subdivided into four irrigated systems of 60 acres each. The remainder of the farm is too hilly to irrigate and is grazed as a less intensive dryland system. Puna Grassland Chicory has been added to increase the mid-summer yield of the dryland acreage. Each of the four irrigated systems runs 120 head of dairy and dairy-cross bulls. These 120 head systems are then further subdivided into six mobs of 20 bulls. This small mob size is due to the use of intact bulls rather than steers.

The New Zealanders have found that mobs of 20 bulls are the most manageable herd size and the group size least likely to fight and ride subdominants in the herd. Dairy bulls are used because of their lower initial purchase price and superior growth characteristics compared to the traditional beef breeds.

Wier has worked hard to make his farm what he calls a "technosystem." By designing a coherent "whole" production system, easy to understand in both its layout and operation, he hoped to be able to sell a turn-key beef production system that could be replicated elsewhere in New Zealand and the world.

Simon Grigg of Burleigh Beef's hardware division Kiwi-tech International explained that the New Zealand beef market is made up of two sectors. One is the traditional "Prime Grade" (same as USDA Select) produced from 1000 lb Angus-cross steers, and the other is manufacturing grade beef from the dairy bulls. This beef is primarily exported to the USA hamburger market. (New Zealand exports approximately 500 million pounds of beef to the USA annually.)

Whereas New Zealand lamb and dairy products bring a price far less than their North American equivalent, New Zealand grassfed beef tracks USA prices pretty closely with NZ Prime selling at roughly the same price as USDA Select and the manufacturing grade bulls selling for a price equivalent to upper Midwest fed Holstein prices.

In New Zealand the slaughter price is the highest price per pound of any class or weight group, whereas, in the USA on the current market it is the lowest. This means the New Zealand grazier is actually working with a higher value per pound of gain than the USA grazier. Also on the manufacturing grade market there is no quality grade, but premiums are paid for heavier than normal carcasses. These factors have made dairy bull grazing the most profitable pastoral enterprise available to New Zealanders after dairying.

Grigg said that Burleigh aims for a slaughter liveweight on their bulls of 1300 lbs at 24 months of age. The Fresian (Holstein) bulls are purchased as 600 lb yearlings from herds of contemporaries to prevent fighting and riding. Care is taken to never mix groups of bulls at any time during their stay at Burleigh.

One 60 acre "system" consists of six ten-acre lanes of grass, 55 feet wide by 880 yards long of permanent two-wire (both hot) fence. The grass is then rationed with temporary polywire and portable posts. Two portable polywires, one in front and one in back, stretch across the entire six lanes and all six mobs of bulls are moved at the same time. Once they reach the end of their lane, they are then driven back over the paddock to start at the other end.

The bulls are shifted every other day as frequent paddock shifts help keep the bulls from becoming bored and thereby helps stop fighting and riding. The grass rest period is varied by altering the length of the paddock rather than the number of paddocks.

Permanent wooden fence posts are placed every 22 yards so that each fence post division will equal one-quarter acre. This clearly marked and known minimum paddock size helps make dry matter budgeting easier to figure. The bulls are rotated on a varying paddock size according to so many posts per subdivision. For example, a one post shift every two days will result in an 80 day rotation. A two post shift will result in a 40 day rotation, and so on. This post system of movement also allows the manager to clearly instruct untrained labor as to how far to move the bulls per shift.

Portable plastic water tanks equipped with high flow valves shoved halfway under the lane wires water two mobs each and are moved with each paddock shift. A unique Wier-designed push-fit "Python" hydrant connector at every second post allows portable waterers to tap into a black polyethylene mainline water hose laid on top of the ground between every other lane.

A liveweight of around 900 lbs per acre is overwintered on this intensive system. Some hay is fed, primarily to prevent bloat on low dry matter, early spring grass. Maximum use is made of compensatory gain. The bulls actually lose weight during the winter due to tight paddock rationing. However, during the spring lush average daily gains of 4.4 lbs per day are common. A 365 day average of 2 pounds a day is achieved despite the winter weight loss.

Grigg said that a high level of pasture management has to be used in the spring to stay on top of pastures and keep them from growing a seedhead and losing quality. During the spring lush the pastures are loaded up with new stock to keep the grazing pressure on the grass high. The number of subdivisions and the rest period are minimized, but continuous grazing is never used at Burleigh. Animals nearing slaughter are given more grass per shift than animals in the growing stage. This allows the near-slaughter animals to "pop" any suppressed gain.

If you can imagine viewing this river plain from the air, it would look like the interior of a piano it is so strung with wire. If traditional gates were used, it would take one all day just to cross a couple of "systems" and tremendously raise the capital cost to build them. Wier solved this problem by inventing a unique flexible fence that allows the interior fences to quickly be driven over by a bottom-skid-equipped four-wheeler or truck. The wire is fastened to the permanent posts with question-mark-shaped spring attachments that allow the wire to give and then retention itself. Flexible round fiberglass posts are used between the permanent wooden posts. Grigg said Wier got this idea of a flexible bounce-back fence from studying a spider's web and so he named it Spider fence.

This unique fencing feature allows one man to shift the cattle and water of a 60-acre "system" in 20 minutes and the entire farm in less than two hours. Also, a gate can be created anywhere just by stepping on the wire and pressing it down. (Insulated boots, of course.)

To cut out a bull from the herd Wier uses a reel of polywire attached to a short fiberglass fishing pole. The polywire has a weight on the end that allows him to cast the wire like a fishing line. Once he has quietly separated the bull from the mob he will cast the polywire line between the one bull and the mob so that it lands over a paddock fence, thereby electrifying the polywire. He then starts to reel in the wire while he walks toward the paddock fence. The bull moves in front of the hot wire to avoid being shocked and is reeled in like a big fish. When the bull reaches the fence, Wier steps on it and allows the bull to hop over.

All paddock shifts are performed by the paddock manager stepping on the front polywire. The bulls are trained to the fence hopping technique. Occasionally, bulls will make a paddock shift by going through a portable scale so their weights can be recorded.

The paddocks are irrigated by a windup reel/traveling gun irrigator using the tractor as a moveable pull post. The irrigator can travel one half of the length of a paddock in 24 hours. It irrigates two lanes at a time. The gun unit hose, of course, has to be laid under the front and rear polywire fences so the entire irrigation unit — tractor, reel wagon, and gun unit — are equipped with upside down U-shaped skids. These pick up the polywire and allow them to drive under it. Bottom mounted skids allow them to drive over the permanent lane fences. Grigg said that no bulls had ever escaped while the irrigator gun unit had the polywire lifted as they are apparently afraid of the water-squirting gun and stay away from it.

While originally designed to be a turn-key, hardware only "technosystem," Grigg and Wier subsequently learned that management-intensive grazing hardware doesn't sell well in North America unless accompanied by the "software" or operator knowledge of how to use it. "A technosystem is a combination of hardware, grass, animals, and the operator. But the skill of the operator is by far the biggest part," Grigg said.

Big Profits from Little Holsteins

Research at the Coastal Plain Branch Experiment Station in Newton, Mississippi, has shown grazing baby Holstein steers to be far more profitable than grazing traditional beef breed stocker cattle during the high price phase of the cattle cycle. Profits per acre in the 1989-90 and 1990-91 seasons from winter grazing baby Holsteins have ranged from between $325 and $552, depending upon the stocking rate used. In contrast, the ten year average profit per acre for beef breed ryegrass stockering in Mississippi during the same period was around $150 an acre.

Demand for Holstein feeder steers is extremely dependent upon the stage of the cattle cycle. Demand is good during the high priced phase when the withdrawl of heifers from the feeder cattle stream dries up the feeder cattle supply. Conversely, there is virtually no demand for them during the liquidation phase of the cycle when traditional beef breed feeder cattle are plentiful.

In the South, the majority of dairy calves are born in September and October. These light weight calves can make exceptionally good use of winter pasture due to their low body maintenance requirements.

"The growth rate of a fall-born baby Holstein steer almost perfectly matches the seasonal growth rate of annual ryegrass pasture," explained station superintendent Bill Brock. "He is extremely small and has minimal feed requirements during the dead of winter when grass growth is low, but as the grass turns it on in the spring, the Holstein does too and consequently very little pasture is wasted."

Gains of 1400 to 2200 Pounds Per Acre

Due to this excellent mesh of animal and grass growth, gains per acre from heavily nitrated Marshall ryegrass continuously grazed with baby Holstein steers have ranged from 1400 to over 2200 lbs of gain per acre. The Newton program starts in early fall with newly born day-old Holstein calves. Due to both market and production advantages, fall calving is recommended for Deep South dairymen.

The baby calves are raised on milk replacer in individual hutches on pasture until they are five weeks old. They are then shifted to small holding traps and fed hay and grain until they are two months old. At two months of age, the calves go on annual ryegrass pastures and are grazed until the following May. Due to poor heat tolerance, Brock recommends the Holsteins be sold to midwestern buyers in late May when ryegrass pastures fade. "Our original idea was to finish these calves on whole shelled corn here in Newton, but such long feeding programs (6 months) don't fit the feed resources of the South as well as a grass program," he said.

He said they have had no problem selling the 650 lb Holstein feeders through local market channels, but warned that graziers in other areas should make sure they have a market before going into a Holstein grazing program.

Death Loss Must Be Kept Low

A key to dairy beef profitability, Brock said, was to keep death loss low on newly purchased day-old calves. "For this program to be profitable, you have to keep your baby calf death loss under 10 percent. A death loss of five percent or so is acceptable."

Assistant dairyman, Joey Murphey, said a critical factor in calf death loss was the protein percentage of the milk replacer used. "It absolutely has to have a minimum of 20 percent fat and 20 percent protein. If the replacer is one percent below 20 percent, you'll start losing calves," Murphey said.

Brock said beef breed stocker graziers need to keep in mind that very few dairy calves have had any colostrum and so are very susceptible to disease. "A day-old dairy calf is an entirely different animal than a six or seven month old beef stocker calf. Your management and health practices have to be superb. You definitely need to use the correct milk replacer and work closely with your vet."

To prevent disease buildup, the calf hutches are moved to a new pasture after each set of calves. The cleansing power of sunlight has proven to be the best pathogen preventer for baby calves. Barns should be avoided.

Calves Start Grazing at 130 Pounds

The Holstein steers at Newton average 130 to 175 lbs at two months of age when they are ready to go solo on late fall ryegrass pasture. Brock said gains on the calves will be slightly less than 1.5 lbs per day during the mid-winter period and this was quite acceptable. "Remember, we want to take as little body weight through the worst of the winter as possible to minimize body maintenance costs and maximize forage use efficiency and profits," he said. If the pasture gets short, he recommends the calves be supplemented with three to four pounds of whole shelled corn a day rather than hay due to their small rumens. "The real beauty of the Holstein steer calf is the way he turns it on in the spring. Our per head spring gains are in excess of three pounds a day. The calves just grow into the spring ryegrass lush and keep up with it. In early May when the ryegrass is at its maximum growth our initial stocking weight per acre has grown from 600 lbs to 2700 lbs at our highest stocking rate. Little to no grass is wasted or has to be cut for hay. It all gets converted into beef."

Brock said the station had experimented with stocking rates and had used an initial November stocking rate of 3.0, 3.6, and 4.5 calves per acre. Calves at the 4.5 stocking rate are supplemented with a 16 percent crude protein corn-soybean meal grain mix at a rate of one percent of live bodyweight daily. Brock said the high stocking rate had produced both the highest and lowest profit per acre ($325.13 and $553.33).

High Stocking Rate is High Risk

Due to its variability and the increased market risk from lower per head gains, Brock said he did not recommend the highest stocking rate to commercial graziers. "The 3.6 stocking rate appears to offer the best of both worlds. A high return per head and a high return per acre," he said. The 3.6 stocking rate has produced a return per acre between $389 and $413 and an average profit per head of around $110.

The pasture used with all stocking rates is a pure stand of Marshall annual ryegrass fertilized with 200 lbs of N. The nitrogen was applied at 70 lbs at planting, 50 lbs in mid-December, and 80 lbs in mid to late February.

"This is not cheap pasture to produce. If you're going to put a lot of weight out there, you've got to put a lot of nitrogen out there," Brock said. Total agronomic cost per acre is around $160 due to the high nitrogen rate, but this produced a pasture cost per pound of gain of only eight to nine cents per pound at the 3.6 stocking rate. Brock said that as dairymen across the country got their cows' breeding schedules in tune with the grass growth cycle, the opportunities for dairy calf stockering would increase. "If the market really wants lean beef," Brock said, "a Holstein can probably produce it cheaper than any other breed. The lower initial cost per head and the much higher return per acre makes dairy beef production a major opportunity for stocker graziers near dairy production areas."

Set Stocking Best with Light Calves

Set stocking has been found to produce a higher average daily gain than rotational grazing with very light calves of less than 350 lbs. Light calves can select a very high quality diet with their small mouths. Grass for light calves should be kept very short, young and leafy. Ryegrass, bluegrass and bermudagrass make excellent forages for light calves as long as they are kept about three inches tall. A variable stocking rate that can keep pasture utilization high throughout the season can be produced with set stocking and a fixed number of calves with buffer grazing.

With buffer grazing a movable electric fence is used to increase or decrease stocking rate by increasing or decreasing the size of the paddock. The buffer area can be cut for hay or silage if it gets too mature.

The goal of grazing management during the green season with stocker cattle is to produce a variable stocking rate and thereby keep the grass sward young and tender but of constant quantity. Buffer grazing is an excellent way for stocker graziers to start practicing managment-intensive grazing with sod-forming grasses like ryegrass, bluegrass and bermudagrass.

Dairy Replacements Offer Greater Profits for Gain Sellers

Graziers can usually double their income per pound of gain by grazing dairy replacements rather than beef steers. The short production life of most North American dairy cows (less than three lactations) keeps grade, pasture-raised, bred dairy replacements in good demand.

Today, even the most die-hard confinement dairymen are realizing that pasture grown replacements tend to have bigger rumens, and therefore greater feed capacity (which makes more milk), than confinement grown heifers. According to New Zealand consultant, Vaughan Jones, research in Britain has shown that the size of a calf's rumen at three or four months of age remained relative for their lifetime. This is why it is so important that calves get onto pasture very early in their lives.

The value of gain for dairy replacements has traditionally been priced at around one dollar per pound. The reason for this is that it costs approximately a dollar a pound to raise a heifer in a "modern" confinement system. Due to these high costs, more and more dairymen are seeing the value in contracting out the growing out of their replacement heifers to lower cost professional graziers. Even pasture-based dairymen have figured out that turning all of their grass into milk is far more profitable than using it to grow their own replacements at home.

In California, some dairymen put their replacements out on grazing contracts for as high as 85 cents per pound and think they are making out like a bandit compared to the cost of raising them at home. Gain rates tend to be higher the lighter and younger the animal is due to the higher level of management required. For calves over 400 lbs, cost of gain payments in the $45 to $55 cwt range are typical. This value of gain is virtually equal to the average sell-buy from non-leveraged beef stocker steer ownership, but of course, gain grazing requires no capital for animal ownership.

A New Industry in the Making

The contract grazing of dairy replacements is a huge industry in New Zealand with highly formalized grazing contracts, periodic weighups, and guaranteed minimum gains and target weights. The current USA productivity goal is for the heifer to breed at 15 months and calve at 24 months of age. To achieve the target weights this breeding schedule calls for requires a relatively high constant rate of gain year around with today's large frame, late maturing North American Holsteins. Due largely to poor winter weight gains, a sizable percentage of North American Holsteins do not calve until they are 28 to 29 months old. Vaughan Jones gives the following minimum weights to aim for in a heifer's life.

Mimimum Weights to Aim For

	Weaning	8 Months	Mating	Before Calving
Holstein Friesian	165	550	750	990
HF X Jersey	155	420	640	880

It is this necessity for keeping animals gaining on schedule that earns the contract replacement dairy heifer grazier his premium price per pound. No doubt as the contract graziering industry matures over the next few years, periodic every other month weighups will become as standard in North America as they are in New Zealand. Weighups will reassure the dairyman customer about the grazier's skill and provide a benchmark for periodic payment to the grazier. In fact, a complete New Zealand-style graziering industry is ripe for development in North America. In such a case, a large master contractor handles the marketing to find dairymen customers; subcontracts out the grazing to smaller graziers; oversees the periodic weighing and reporting of gains; schedules the trucking; and handles the billing and paying of gain money.

An Alternative to Contract Grazing

An alternative to contract grazing is buying newborn heifer calves from local dairymen, growing them out and selling them as bred replacements at 18 to 20 months of age before their second and most expensive winter. Most graziers I talked with report an average value of gain from this of around $84 cwt after marketing costs, price rollback, non-breeders, and death loss on the baby calves are figured. This is far better than the $40 to $50 cwt value of gain typical with purchased beef stocker steers; however, most of the replacement heifer graziers I talked with would prefer a grazing contract over owning the heifers themselves. Keep in mind a pregnant heifer has a ticking clock inside her. The heifer must be sold before she calves or you are in the dairy business. Some dairymen have been known to take advantage of replacement heifer graziers with a herd of soon-to-calve heifers.

Also, the long production cycle produces cash flow problems for graziers. Beef oriented bankers who are used to "touching" their money every six to twelve months (paying off the old loan and then writing a new one) are uncomfortable with the 18 to 20 months their money is tied up in a dairy replacement. Another risk is that a certain number of heifers will be unable to breed and will have to be sold on the beef market for whatever they will bring.

The increase in seasonal dairying is having an impact upon replacement dairy heifer demand. It is important that the heifers calve slightly later than the older cows so they can become accustomed to the milking routine before they have to do it for real. Northern dairymen increasingly want their replacements bred to calve in March/April while Southerners prefer September/October and/or December/January breeding. Having customers in both the North and South helps with heifers who fail to breed on schedule.

In the South, many dairy producers traditionally use natural service beef breed bulls and buy in all replacements. Many contract graziers also use easy calving beef breed bulls to avoid the physical danger of highly temperamental dairy bulls or the hassle of AI. Due to the large supply of this type of calves, the South and Southwest has an active market for lightweight, beef/dairy cross calves.

Death losses on calves from first calf heifers can be extremely high. First calf heifers frequently do not produce enough colostrum to adequately protect their calves' health and their calves should also receive colostrum milk from older cows.

Three Management Phases

I've observed that when any livestock production program extends over six months in length, it results in the animal changing ownership several times as graziers start to specialize in whatever phase best suits their talents and climate. Replacement heifer rearing lends itself particularly well to North-South partnerships. The dairy replacement business is broken into three major management phases:

1. Birth to weaning at six to nine weeks of age.
2. Weaning to breeding at 15 months of age.
3. Breeding to calving at 24 months.

Women Best at Calf Rearing

The birth to weaning period is very management intensive but compensates for this by producing the highest value of gain. Values of gain in excess of a dollar per pound are increasingly attracting small acreage, "specialist" graziers who sell or contract out the weaned calves to others for continued growing. The Paul McCarville whole-milk-on-pasture method of raising baby calves has recently tremendously increased the profitability of this pre-weaning period by lowering death loss, labor, feed and capital costs.

Vaughan Jones believes that calf rearing is best done by women. "Calves respond to thoughtfulness and tender loving care, and women are usually more gentle and thorough," he said.

If the weather is nice, baby dairy calves should be out on pasture within a week of birth. The overfeeding of milk and meal will result in calves with poorly developed rumens that "crash" after weaning. The calves must be given an incentive to graze. These calves require very short and tender grass to gain well. To see what a pasture for a week old dairy calf should look like take a look at your front lawn. Grass in baby calf pastures should never be over four inches in height. To see what a pasture for a baby calf should feel like take your shoes off and walk across it barefoot.

If you use a mower to shorten your pastures, you will find this to be a very unpleasant experience as mowing produces a prickly stubble. The way it feels on the bottom of your tender foot is the way it feels on the calf's tender nose. Baby calf pastures need to be kept in condition with grazing rather than mowing. Keep in mind the microflora which develop in calves' rumens are passed down to them by adult cattle. This means baby calf pastures have to be periodically grazed by cattle with functioning rumens. If the calves quickly develop a potbelly after weaning, it is usually a sign of a mis-functioning rumen.

Profit potentials for this specialty are probably the highest in California and Florida where large, industrialized dairies traditionally dump their newborn calves for whatever they will bring. As a result, dairy calves of both sexes are often a third or less of the upper Midwest price. Profits in excess of $200 per heifer from a six week ownership are said to be relatively common in these areas of the country.

New Zealand dairy producers typically graze their spring-born replacement calves through their first summer on their own farm and then put them out on contract grazing until breeding at 15 months or until just prior to calving at 24 months. (On many New Zealand dairies all animals are wintered off-farm on grazing contracts. This allows the minimal winter grass growth to accumulate and prevents pasture pugging damage.)

Between Weaning and Breeding

Weaned calves are either set-stocked two to three per paddock and the lactating cows rotated through the paddocks or the heifers are grazed as a leader herd in front of the lactating cows. Set-stocked young calves are very territorial and do not try to follow the cows out of their home paddock. If rotational grazing is used, mob sizes should not exceed 50 calves.

In Ireland, the spring-born weaned calves are used as a follower herd behind the lactating cows. The Irish believe that gains in excess of 1.5 lbs per day prior to sexual maturity retard the development of the heifer's mammary glands and hurt subsequent milk production. The Irish are careful to control the residual left by the lactating cows in each paddock to prevent extremely high gains on the follower replacements. Higher rates of daily gain are allowed once the heifer reaches sexual maturity around eight months of age.

After the heifers are bred at 15 months, they are treated the same as dry cows and are used as a paddock clean up crew. Therefore, a spring calving Irish dairy would have a leader-follower rotation of lactating cows; followed by new crop heifer calves; followed by old crop bred heifers.

No doubt the biggest opportunity for North American contract graziers is the period between weaning and breeding. This is pretty typical of beef stocker grazing with the exception that you are starting out with a less than 200 lb calf. At this weight, internal parasitism can be a major problem and every effort should be made to keep both the calves and the pastures parasite free.

Newly renovated pastures, or croplands being shifted to pasture, are ideal for new crop heifers as they are parasite free and the grasses are young and tender. Also, the light bodyweight of the new crop calves helps prevent bogging damage on these sodless swards as well, so it is good management for the pastures as well.

Pastures that have been previously grazed with sheep, cut for silage or hayed are also good new crop calf pastures. Prepared seedbed pastures of winter annuals such as rye, annual ryegrass, oats and wheat are also parasite free.

Gains per acre with calves less than 400 lbs can often be quite spectacular (1000 lbs or more) due to their low body maintenance requirements. As the calves age and gain weight, management intensity drops but so does the profit potential.

After the heifers are bred, care should be taken to make sure they are not allowed to get fat on pasture and should be kept in medium flesh. This will help prevent subsequent calving problems. Also, cows that are fat at calving are susceptible to milk fever and are harder to breed back than those in medium flesh. These bred heifers are excellent to use as follower stock in an Irish-style, leader-follower rotation. Whatever method you use, grazing will have to be restricted in some manner to prevent excessive fattening.

Dairy Heifers Plus Beef May Offer Best Pasture Utilization

Eric and Bridget Anderson's, 2800 acre Puketotara Partnership is always one of the favorite stops on our New Zealand visits. Once when we visited, bull beef for the American hamburger market was the number one dollar earner on the ranch. Another time it was replacement dairy heifers. In the future, Eric said it may be sheep or deer. The only constant in graziering is your ability to grow quality pasture. And, as Eric will tell you, neither livestock nor land ownership is necessary, nor possibly, even desirable.

In 1989 when Eric was only 28, he leased the semi-rundown ranch, which had gone bankrupt under a previous ownership and had been bought by wealthy off-farm investors. The annual lease runs between 3 and 6 percent of appraised value. (The investors are betting on land appreciation for a higher eventual return.)

The ranch has only 360 flat acres (150 ha) with most of the ranch classified as undulating to steep and some 600 acres are not grazeable at all. It originally had only 55 permanent paddocks, had not been fertilized (lime and phosphate) in several years and had suffered below maintenance fertilizer applications for many years prior to the bankruptcy. Still, in his first year on the property, Eric was able to contract graze 1400 heifers, 200 steers, 350 cows, 830 bulls and 3000 ewes for other farmers.

He took the profits from these contracts and invested it in paddock subdivision and fertilization. He subdivided the property into 165 permanent paddocks, put in a water system with an in-line mineral dispenser, two stock yards, and initially fertilized each acre with 350 lbs of phosphate and a ton and a half of lime. "We saw a 100 percent return on the money we spent on pasture subdivision over the next two years," he said.

Today, Eric is grazing 1250 dairy replacements, 500 two-year-old hamburger bull beeves, 200 beef breeding cows, 90 yearling beef steers, 60 beef heifers and sheep numbers are down to only 100 ewes, kept as Eric said "for lawn maintenance." However, the sheep shearing shed and yards are kept in good repair and sheep could be back in a flash if prices warranted them. One thing Eric makes very clear is that he is not married to any particular type or class of ruminant and can change enterprises quickly if the market warrants it.

He has one 70 HP tractor, a silage feed-out wagon, a pasture topper (rotary mower), four ATVs, and three employees. Each employee carries a cell phone at all times. All pasture silage is cut, carried and packed by outside contractors. 400 kg per ha (400 lbs per acre) of Sechura reactive rock phosphate are applied annually. Supplemental selenium is provided through the water supply.

Eric has seen a tremendous decrease in bloat problems since he started using the slow-release reactive rock phosphate instead of superphosphate. Nitrogen is used sparingly as a feed supplement to meet grass growth deficits.

His management goal is an effective farm surplus (return over all costs before interest, salaries and wages) of $600 per hectare ($250 per acre) from the entire property or around $500,000 per year. Currently contract dairy heifer grazing returns US$1050 per hectare ($437 per acre), bull beef $750 per hectare ($312.50 per acre) and beef cow-calf approximately $500 per hectare ($208 per acre).

Contract Dairy Heifers

Erik grazes dairy heifers on 12-month contracts on a weight gain payment basis. The 1250 heifers are stocked at 2.8 heifers/ha (1.2 per acre) in mobs of 50 in winter and 150 in summer. Their average incoming weights are 220 kg (485 lbs) on May first going out the following April at an average of 490 kg (1080 lbs). May is the Northern Hemisphere monthly equivalent to October.

Clients are drawn from the nearby dairy-dense Waikato region. Currently he has 18 dairy replacement clients. His smallest client has 30 heifers on contract and his largest 180.

Erik receives $4.85 a week — paid monthly — for grazing Friesian and Friesian/Jersey cross heifers and $7 a week for Jersey heifers. The higher price for the Jerseys is because they can't be mixed with the Friesians and therefore require special attention.

Heifers are weighed every two months. The owner receives a report on each heifer showing her latest weight, previous weight and her average daily gain. Consistency of performance is very important in the replacement heifer business. Friesians will average 1 kilogram per day (2.2 lbs) on pasture and slightly less on pasture silage. Death loss averages two heifers a year but no count guarantees are offered. "It's a reputation business," he said. "A trust thing.

If I don't do a good job, I'm out of business."

Erik charges $8 per cow to get her up for A-I breeding. The heifer owner must provide the semen and A-I technician. Angus, Jersey, or Hereford bulls are provided for natural breeding at no cost. Hereford is the most popular breed choice because the black, white-face cow is the most in demand by local beef farmers. These bulls are all bought in as yearlings and are resold to beef farmers as two year olds. Because these young bulls increase in price as they grow and age, he can offer the in-pasture breeding service free. Erik offers no guarantee on breeding but reports 90 to 94 percent go home bred. He likes contract dairy grazing because it requires no capital outlay and provides a monthly cash flow. However, because the number of animals is set and performance must be high, an all-dairy heifer-grazing program does not offer enough flexibility in stocking rate to match the seasonal grass growth curve.

Manufacturing Bull Beef

Erik buys in 18-month-old Friesian bulls in April/May (October/November in the Northern Hemisphere) at 410 kg (900 lbs) and sends them to slaughter in the December/January (June/July) time period at around 660 kg (1450 lbs). These bulls are contracted with the slaughter plant at purchase to avoid excessive price rollback. Their primary benefit is they offer a quick way to destock and closely match summer pasture growth to summer rainfall. "We farm to the summer dry here. That's our major flatspot," he said. "We want all of our bulls gone by first of January (July in North America)."

Beef Cow-Calf

Erik's beef cow-calf herd is another tool used to match the feed demand/growth curve. The cows are timed to calve in sync with the start of the spring lush. The calves are short-weaned prior to the onset of the summer dry. The beef cows are contract grazed off the farm with rowcrop farmers during the winter.

"The beauty of a beef cow is that she tremendously increases her pasture consumption at calving but then her feed demand drops like a brick after weaning. We see our beef cows as walking pasture toppers," he said. The calves are retained after weaning and sold at 15 months of age as stocker cattle prior to the onset of the summer dry. This adds another flexible class of stock to the ranch. All beef replacements are purchased.

Erik said a grazing property had to have a mix of classes and types of livestock to best match its pasture growth curve. A PastureGauge is used to measure each paddock and the amount of dry matter on hand is kept up with on computer. "The feed budget is more important to a grazier than the financial budget," he said.

For example, Erik feeds his spring-saved pasture silage in the late summer and autumn and allows his pasture to stockpile for winter use. Silage is fed until the minimum required pasture cover of 2500 kg per ha (2500 lbs per acre) for the winter grazing period is reached. The ranch makes 1500 metric tons of pasture silage a year in five separate stacks (clamps) of 500 tons each. One 500-ton clamp is always held in reserve for any unexpected weather event.

The 165 permanent paddocks are subdivided into 480 temporary paddocks with one-wire electric fences. The cattle only receive water every third day during the winter. No silage or supplements are fed at all in the winter.

To better utilize his spare time he has recently leased two dairies with 450 cows and a 5000 acre hill country sheep station. "You've got to take the opportunities as they come," he said. "I figure a man has pretty much got to have accomplished all he wants in life by the time he is 45 or 50."

And does he plan to ever purchase Puketotara? Erik shakes his head. "I've seen too many farmers hell-bent to own a farm just for the sake of owning it and they get over extended. It's really all about cash flow. The goal is to be your own boss, not to own a farm."

Mixed Species Grazing

Dr. A. N. Nicol of Lincoln University in Canterbury, New Zealand told me that mixed species grazing not only increased utilization of the forage sward, but through the "magic" of synergism actually resulted in a greater output than was possible from the same liveweight of any one species.

Nicol said there are very few sheep-only farms in New Zealand. The typical New Zealand hill farm will be 55% sheep and 45% beef cattle (by liveweight). On a hill farm, beef cattle naturally prefer to graze the bottoms and shady slopes and sheep the hill-sides.

Internal parasites are major problems for sheep and goats in humid climates, and the presence of cattle help break the parasite cycle. With one exception, cattle and sheep internal parasites are different. Ingestion of sheep parasites by cattle kills them and vice-versa. However, sheep and goat parasites are the same and present problems when grazed together in humid climates.

He said mixed species grazing required the correct matching of the stocking rate percentage of the various species to the forage resource. There should be as little dietary overlap as possible for synergism. For example, beef cattle and goats are almost totally complementary with very little dietary overlap. Goats can increase grass pasture quality by grazing off stems and seedheads and make an ideal companion species for beef stocker cattle. Goats also hate clover.

Because sheep will graze closer to cattle dung pats than cattle will, adding a small scavenger follower flock of sheep to a beef cattle operation can increase total product output by 5 to 10%. He said grass growing in beef dung pats was attractive to sheep for about six weeks.

Conversely, in a sheep dominant system adding a few yearling stocker cattle will improve the forage resource as the cattle can consume the vegetative stems and seedheads that will be rejected by the much more selective sheep. In a ewe-lamb situation, seasonal stocker cattle are better than cow-calf as the cattle can be removed after the fast growth of spring and late summer. This will allow a surplus of grass for ewe flushing in the fall. Lambs should always be the first grazers in a mixed species system to allow them to finish before winter.

Sheep performance is almost always increased by the

presence of beef cattle. However, he warned, the reverse is not always true.

Because sheep have smaller mouths and can select a more nutritious diet than cattle, in a set-stocked situation stocker performance will be lessened by the addition of sheep if the sheep numbers exceed 25% of the total stocking rate. However, using rotational grazing, the gain advantage shifts back to the beef cattle.

He said in humid climates rotational grazing causes beef liveweight gain to increase and sheep liveweight gain to decrease. Under a short duration graze such as is used with rotational grazing, the beef animal's large mouth becomes a competitive advantage. With long duration grazing the sheep's small mouth has the advantage. An excellent grazing management alternative to overcome this dilemma is to set-stock the sheep and rotationally graze the cattle through the sheep paddocks. This keeps the grass highly vegetative.

He said the rule of thumb to use with cattle/sheep systems is "Got an excess of forage? Then a sheep/cattle combination is a win-win situation. Got a shortage of forage? Then the sheep will win and the cattle will lose."

Pasture Profits

* The grazing of replacement dairy heifers offers a higher profit potential, both from contract grazing and ownership, than beef steers in most years and particularly at the peak of the cycle.

* As with all gain grazing opportunities, it is a reputation business that requires you to start small and build your business a client at a time.

* Mixed species grazing not only increases utilization of the forage sward but actually results in a greater output than is possible using the same liveweight of any one species.

Chapter 14

Stocker Forage Tips

"A wise man is glad to be instructed, but a self-sufficient fool falls flat on his face."Proverbs 10:8

One of the most rewarding parts of stocker grazing in the humid areas of North America is the ability to greatly extend the "natural" grazing season through the use of various plants and grazing techniques. The grazing season can be extended by starting earlier in the spring and going later in the fall. With the use of warm-season plants and/or the use of brassicas such as rape, the summer slump in grass production can also be greatly alleviated.

The following forage growth chart is for Pennsylvania. Work with your extension forage agronomist and identify plants that can help even out your annual forage growth curve. Always experiment with new plants on a small scale first. Each species and variety has a "learning curve" attached to it. Unfortunately, very little research has so far been done with brassicas and stocker cattle in North America, but results with dairy cattle and sheep appear to warrant a better look at these plants. Later in this chapter you will find some of the more popular stocker forages in North America. Keep in mind most of these are best used in conjunction with other forages to extend or balance the forage production curve.

Forage Growth Curve

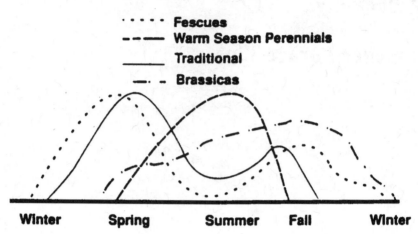

Forage growth by season of the year showing fescue (. . . . .),
traditional cool-season (_____), warm-season perennial (_ _ _ _),
and brassica (_. _ .._ .) growth.

Source: Penn State

Nitrogen Fertilizer Can Help Balance the Forage Curve

Periodic applications of nitrogen can help flatten the huge
spring growth bias of cool-season grasses. In **Grass Productivity**,
Andre Voisin recommended that a grazier's yearly nitrogen application on cool-season grasses be split into one-sixth of the total just
before spring growth begins, one-third at the end of the spring
grass lush, and one-half in the autumn to build a stockpile of grass
for subsequent wintergrazing. (For the mathematical formula on
determining your total annual nitrogen application see Chapter 6 of
Grass Productivity.)

Voisin found that with timely nitrogen applications he could
increase the grass growth season by nearly one month a year,
diminish the effect of the summer grass growth slump, and even out
the roller coaster growth curve of cool-season pastures.

Cool-season grasses are agronomically able to grow at
temperatures lower than clover rhizobia, and soil bacteria can start
nitrogen production. By applying nitrogen in the late winter, he

enabled the grass to begin its growing season before that of the natural soil nitrogen production cycle.

These late winter applications are particularly important if the fall was droughty and the grass was carried into the winter very short. These short-grazed autumn pastures will be very slow waking up in the spring due to low root reserves for growth.

Voisin noted that with a late winter nitrogen application he could begin grazing when the cherry trees were just beginning to bud, but if he didn't apply nitrogen, the grazing season began when the cherry trees were blossoming.

He found a cool-season grass's slowing growth due to the onset of summer heat could be sped up with an application of nitrogen as well. This late spring application greatly diminished the effect of the summer slump on grass growth.

Care must be taken with spring nitrogen applications so that they are not made so late that they actually worsen the spring growth surplus. He found that one unit of nitrogen would grow 5.30 units of grass in the spring, but only two units in the fall.

Voisin recommended that in a ten paddock dairy cell only the first six should receive a late winter application of nitrogen. He recommended the first three paddocks should receive 134 lbs of N and the next three should receive 89 lbs of N. Two of the four paddocks would receive no nitrogen and would be set aside for haying or silaging. Two would receive no nitrogen but would be grazed in the spring rotation.

This staggered rate of application helps give a "cut" to the spring rotation and helps prevent clover bloat by increasing the grass percentage.

Each year the rotation should start with a different paddock to prevent favoring certain forage species over others.

Cattle that have been wintered on hay and feed should not be allowed to graze but two hours on the first day. Their grazing period can be gradually increased each day.

If the cattle are not gradually allowed to build their rumen microbia for spring pasture and legumes, severe bloat and other problems can occur.

Stockers in the Subtropics

Mid-September to mid-November is the worst pasture production flatspot in the sub-tropic pasture region of the Deep South. Due to a summer evaporation rate much higher than cool-season perennials can withstand, warm-season tropical perennials such as bermuda are used to form a pasture base. However, these heat-loving tropical species go dormant during the five cool months of the year. This winter gap forces Southern graziers to rely heavily upon cool-season annuals (wheat, oats, cereal rye and annual ryegrass, solely and in combination with one another) for winter and early spring grazing.

Most Deep South stocker programs are built around high quality cool-season annuals. These annuals produce a very high quality pasture for 150 to 200 days in the cool winter and early spring. Thanks to exceptional stocker growth rates, these nominally "high cost" annual pastures are actually very economical in the sub-tropic region where an actively growing winter pasture is possible.

Many graziers use a combination of prepared seedbed and overseeded (cool-season annuals drilled or spun onto short-grazed tropical grass) pastures. Providing a seamless transition from the tropical grasses to the cool-season grasses is an annual test of a grazier's skill.

The tropical grass production curve peaks in mid-July. The daily growth rate then falls precipitously as day-length shortens. Trying to bridge the fall gap by planting cool-season annuals earlier encourages insect attack (Armyworms) and drought stand failure in prepared seedbed pastures.

Also, early season overseeding of the tropical grasses can be stymied by warm, wet falls that keep the tropical grass growth rate high enough to shade out the cool-season seedlings and thereby create a stand failure. Consequently, many Southern graziers have just resigned themselves to later planting, a "fall flatspot" and a four- to six-week period of stored forage feeding. One who hasn't is East Texas grazier, Steve Roth. While Steve has perfected his program with dairy cows it is perfectly transferrable to stocker cattle.

Steve, his wife Pam, and their three sons graze 1200 dairy cows on 600 acres of Coastal bermuda (seasonally overseeded with cereal rye and annual ryegrass) near Grand Saline, some 60 miles (100 kilometers) east of Dallas. The Roths traditionally drilled cereal rye into their Coastal in October and then overseeded (with a cyclone seeder) with annual ryegrass in December and January.

After several years of trying to plant his rye earlier and earlier, he began to notice that the thickest stand of December grass was always from the paddock planted last. He said this caused him to rethink his whole planting program.

"I was using a forage contractor and drilling the whole farm in October. Once you had drilled in the rye, you felt you had to stay off of it even though there was still lots of grazeable Coastal. It just bothered me that we were feeding all this hay with so much grass still around.

"Now we don't start drilling rye until around the first of November or whenever our first frost hits and puts the Coastal to sleep for the winter. We have bought our own pasture drill and now drill each paddock after the cows leave it. Normally it has grown a good stand of rye when we return in 30 to 45 days."

Rotation Length Only 10 Days in Summer. Each Roth dairy (two are currently in operation and another is planned in the near future) is split into 40 permanent paddocks. These are then further subdivided with temporary electric fence to provide a variable rotation length. Rotation lengths are as short as 10 days during the summer with the fast growing Coastal and are only 15 days in the spring with the annual ryegrass. In the winter, they average 35 to 45 days.

For graziers unfamiliar with the aggressive growth pattern of tropical forages, consider this: In July, each acre of Coastal requires four Holstein/Jersey crosses to keep up with the grass growth. To create this high stocking rate, one third of the farm is dropped from the rotation and hayed by forage contractors. In wet summers, it may be necessary to hay the whole farm to maintain grass quality. If so the farm is hayed in thirds.

Steve now uses this aggressive growth pattern to stockpile

paddocks rather than hay them. He starts this around mid-September. He said he estimates he can grow a ton to a ton and half of dry matter between mid-September and the onset of frost in early November. This stockpiled pasture is rationed out with portable electric fences to dry cows from October to early December. In early October, a forage test showed this Coastal to test between 16 and 18 percent protein. "The cows do real good on it," he said.

Stockpiling Program Saves $20,000 in Hay. Roth figures this new stockpiling program saves at least a month of hay feeding. If you figure hay is worth $60 a ton, this totals over $20,000 for 1200 cows. As of mid-November 1996, he had fed no hay at all to get his herd through the Southern "flat spot."

Making the Most from WINTER ANNUAL Pastures

Prepared seedbed cool-season annual pastures (wheat, rye, oats, and annual ryegrass) are expensive, with an out-of-pocket cost of between $80 and $120 an acre, depending primarily upon the supplemental nitrogen level used.

However, these winter annual pastures have the potential to produce high rates of gain, both on a per head basis and a per acre basis. As a result, they can actually be very cost-effective when used with growing animals or lactating dairy cows.

One of the most common uses for winter annual pastures in the South and California is to grow out lightweight stocker cattle. At the Brown Loam Experiment Station, stocker grazing on these annual winter pastures has been a part of the station's research program for over 40 years. This on-going research has shown the enterprise to almost always be profitable.

Per acre returns over the last ten years from the Brown Loam stocker program have ranged from a profit of $179 an acre to a loss of one dollar an acre. The ten year average is $80 an acre.

Gains per acre have averaged in the 600 lbs to 700 lbs range. (If this amount of gain had been sold for 35 cents a pound — the going rate for winter pasture gain in Mississippi — the return per acre would have been almost exactly the same or slightly higher than the ten year profit average from animal ownership.) These variations are primarily caused by varying buy-sell margins between light calves and heavy feeder steers, and have ranged from a minus $27 cwt to a minus $2 cwt. The station has recorded no positive margins in the last 10 years.

A typical calf put into the wintergrazing program weighs about 350 lbs to 400 lbs and will gain about two pounds a day for around 180 days (depending upon the availability of fall rain to get the pasture up) on high-quality winter annual pasture.

A steer calf will about double his weight in the six months the annual pastures last.

At Brown Loam a stocking rate of 1.5 calves per acre is used with continuous grazing. Over the last ten years, supplemental feeding (hay) has averaged about 20 days per winter using this stocking rate.

Brown Loam's research has also shown that it is better to never remove cattle from winter pasture longer than 48 hours at a time, as this will cause rumen upset and the cattle will lose weight. Cattle should have at least some access to the pasture every other day, if at all possible, or be fed small amounts of molasses or molasses blocks daily.

Grazing winter annuals into the ground before starting hay supplementation results in both a slow recovering pasture and a low average daily gain on the calves. Feeding hay earlier usually means feeding less hay over all.

Fall Weight Loss. Research at LSU has shown that cattle turned out exclusively on winter annuals in the fall will lose weight for two weeks and will not regain their original turn-out weight for a month. This is because the rumen bacteria needed to digest cool-season annuals is different from that needed to digest warm-season perennials.

This weight loss can be avoided by turning the cattle on the winter annuals for only a few hours each day at the start of the grazing season, or by feeding a small amount of molasses (liquid or block) for two weeks prior to turning the calves out on the winter annuals.

For many years a combination of wheat and annual ryegrass

was used with Brown Loam's stocker program. The wheat being more cold tolerant gave a better forage distribution than annual ryegrass alone. However, with the advent of Marshall ryegrass, a cold-tolerant variety, research has shown a higher profit per acre from planting pure stands of Marshall ryegrass than from the wheat/ryegrass mix. ($102 versus $90).

Research has shown that higher stocking rates can be achieved with higher levels of nitrogen, but Brown Loam has found that lower stocking rates with lower levels of nitrogen actually produced a higher net profit.

Pastures of Marshall ryegrass and Ball annual clover with a low level of supplemental fall and early spring nitrogen has consistently shown the highest profit per acre.

Gains May Fall. Management-intensive grazing can also produce a higher gain per acre by allowing a slightly higher stocking rate, but care must be taken to make sure that average daily gain is not reduced by poor grazing management. Due to the negative buy-sell margin that exists on stocker steers, a large gain per head is necessary to overcome this deficit. Pastures have to be managed so as to produce a high daily gain per head.

The higher gains per acre may make rotational grazing more attractive to graziers selling gain than those owning the cattle. However as their research with continuous grazing has also shown, higher stocking rates with higher nitrogen rates and rotational grazing have produced a lower net profit per acre. The trick is to get the optimum offtake per dollar invested while still producing a reasonable average daily gain.

Sacrifice areas (tightly sodded bermudagrass pastures or deep sawdust small traps) must be provided for with rotational grazing on winter annuals in the South. This prevents heavy bogging damage during rainy weather.

Marshall ryegrass, Segrest ball clover, and subterranean clover are excellent natural reseeders and can produce good pasture with just a late summer disking, although they are somewhat later than seeded pastures. However, Brown Loam has found it has been more profitable for them to mechanically seed their pastures each year than take a chance on the late pasture that often occurs with natural reseeding.

Bigbee berseem has performed poorly at Brown Loam due to a series of extremely cold winters and the use of continuous grazing. Due to its upright growth habit, Bigbee berseem should be rotationally grazed.

Brown Loam's recommended wintergrazing program involves a summer fallow to rid the pasture of weeds and allow the soil to absorb rain. However, erosion can be a major problem with this program.

Research in the 1950s showed that sod-seeding ryegrass into bermudagrass reduced yields by one-third. In 1989, two studies were begun to see if improvements in planting equipment and herbicides could improve this.

Late Pasture with Overseeding. The bermudagrass was hayed in mid-October and ryegrass was drilled in with no herbicides on one plot, and hayed and sprayed with Paraquat on the other before drilling in September. The first treatment did not produce a grazeable stand until February 14th. The herbicide-treated pasture did not allow grazing until January 16th. Both of these pastures were considerably behind the plowed and fallowed pasture that had a grazeable stand on November 11th.

Another study was made of planting ryegrass into the crabgrass and annual weeds, which naturally volunteer in the summer following annual ryegrass without tillage. Removing the crabgrass by haying appeared to hold a lot of promise, according to the Brown Loam researchers, but spraying the grass and weeds with Paraquat resulted in an outbreak of insects that destroyed the ryegrass stand.

FUNGUS-FREE FESCUE

It appears the endophyte in fescue, that results in the animals being unable to dissipate body heat, actually helps fescue pastures survive in hotter and drier climates. Poor stand persistence across the South for the new varieties of fungus-free fescue has sent University of Georgia researchers back to the drawing boards.

Agronomist Nick Hill and plant geneticist Wayne Parrot, said scientists today believe the endophyte's relationship with its host is more symbiotic than parasitic. Parrot said the endophyte produces alkaloids, which helps the fescue survive environmental stresses such as drought and attacks by insects. Some examples of alkaloids are nicotine, quinine, cocaine and morphine.

Endophytes also infect timothy, bluegrass, wheat, wheatgrass, rye and ryegrass, orchardgrass, fescue, bermuda as well as bamboo, rice, and sugarcane.

A removal of the endophyte in perennial ryegrass in New Zealand resulted in similar poor stand persistence. In contrast, Ellett, the world's most successful variety of perennial ryegrass, is very high in endophyte.

There are apparently three alkaloids in fescue and only one of these affects animal performance. The trick for plant scientists is going to be to remove the one bad one and reinsert the other two that help the plant survive. Parrot said there are now biotechnology techniques that can make this possible.

The Clover Solution for Toxic Fescue. Research by the Kerr Foundation for Sustainable Agriculture and Oklahoma State University has indicated that adding clover to infected fescue is the most profitable solution for wintergrazing stocker steers.

In the Kerr study, Angus steers on a November to May grazing season lost $52.72 on endophyte-infected fescue versus a profit of $96.68 a head for endophyte-infected fescue interseeded with clover. The profit was $61.89 for endophyte-free fescue.

This study did not take into account the high cost of destroying old fescue stands ($125 - $130 an acre), nor the difficulty in establishing and maintaining fungus-free fescue stands, nor its much lower resistance to heat, insects and disease. Interestingly, Brahman X Angus steers appeared to be much less susceptible to the infected fescue and were able to turn a profit of $99.37 a head on fescue. Animal performance was lowered, however, as Brahman X Angus produced a profit of $170.63 on clover/fescue and $110.67 on endophyte-free fescue.

Simmental X Brahman-Angus steers also lost money on the

infected fescue (minus $35.02) but made $96.98 on clover/fescue and $58.78 on endophyte-free fescue. The loss on the Angus and Simmental X Brahman-Angus steers were the result of low average daily gains that prevented enough gain per head being produced to offset the minus $23 rollback in purchase versus sale price.

Other observations from this study were:

☆ Cattle on high-endophyte pastures started showing fescue toxicity symptoms around late March to early April.

☆ Brahman X Angus cattle did not show fescue toxicity signs on high-endophyte pastures. (The researchers noted Brahman X Angus cows frequently lost the switch of their tails when grazing high-endophyte pastures.)

☆ Cattle on low-endophyte forage grazed longer and apparently consumed more than cattle on high-endophyte forage once fescue toxicity started to be a factor.

☆ Cattle on low endophyte pastures did not avoid grazing fescue stems nearly as much as cattle grazing high-endophyte fescue, whether it was interseeded with clover or not.

☆ Cattle removed from high-endophyte fescue pastures in May and placed on ryegrass pastures made exceptional compensatory gains (in excess of four pounds per day) compared to cattle grazing low-endophyte or interseeded fescue.

☆ Cattle from high-endophyte pastures compensated for some of their pasture losses when placed in a feedlot situation.

Please note this was a November to May grazing season. Grazing endophyte-infected fescue past May or whenever the interseeded clovers fade will result in very poor animal performance. Grazing endophyte-free fescue past May frequently results in stand loss. In the South, graziers should confine the primary use of fescue to the cooler months and use warm-season grasses, alfalfa or brassicas as buffers during the warm weather months.

BIRDSFOOT TREFOIL

Birdsfoot trefoil is a perennial legume with the advantages of tolerance to soil acidity and drought, resistance to the alfalfa weevil, natural reseeding from hard seed and a high quality forage that does not cause bloat in livestock. Quite common as a stocker forage in the Midwest, Birdsfoot trefoil has not generally been regarded as a legume that could be grown in the Southern states. However, several culitvars have been found to be adapted to North Georgia. "Fergus" is adapted to the mountains and "Fergus," "AU Dewey" and "GA 1" are adapted to the upper Piedmont region of the South.

Seeding Birdsfoot trefoil in tall fescue sod was most successful when a chemical grass suppressant was applied and the seed planted in rows with a sod-seeder. Broadcast sod seeding was less successful but was improved with light disking. Birdsfoot trefoil maintained its high quality in summer even when stockpiled. In vitro digestibility and crude protein declined less during summer in Birdsfoot trefoil than in alfalfa. Average daily gain of steers grazing "GA 1" trefoil was high, 1.74 lbs for the season. Stocking rate was lower for Birdsfoot trefoil than for alfalfa pastures.

Trefoil Can't Take Bermudagrass. At Eastonton, Georgia, Birdsfoot trefoil pastures were invaded by bermudagrass after their third year of rotational grazing, indicating its lack of adaptation wherever this grass is present. Farther north at Calhoun, "Fergus" birdsfoot trefoil provided high daily gains of over two pounds a day and persisted well in pastures. The research indicates that with good management, this legume can provide high quality, bloat-free forage for good animal performance at least as far south as northern Georgia.

BERMUDA

Coastal and the improved bermudas are not generally thought of as "quality" grasses. The poor animal performance commonly seen with bermuda is a combination of using overly mature forages and hot weather's effect on the animal's grazing behavior. Under cool weather and immature grass conditions, animal response to these tropical grasses can approach that of cool-season species. Bermuda and other improved tropical grasses mature rapidly, and requires a high level of grazing management to be kept of stocker quality.

The secret is to keep the grass short (four to five inches in height) and growing very rapidly. This rapid growth is provided by frequent small amounts of nitrogen fertilizer. Steve Roth applies 40 to 50 lbs of actual N (120 lbs to 150 lbs of Ammonium nitrate) to each acre every 28 days. This high level of fertilization allows (requires) him to graze two cows per acre nearly year around. Even at this high stocking rate, half of the farm will have to be dropped from the rotation and hayed to keep the grass short and tender.

Coastal bermudagrass will produce two to three times as much steer gain per acre as cool-season annuals in the South, yet this grass is seldom used for stocker grazing. This is primarily due to the lack of knowledge of how to graze bermudagrass effectively and, equally important, how to utilize it most economically.

How To Graze It. It takes a fall-planted, cool-season annual six to seven months from the time it starts to initiate growth until it begins to joint, produce stems and mature. The same change from growth initiation and immaturity to stem growth and maturity in bermudagrass happens in about 30 days. The difference between managing bermudagrass and ryegrass is similar to the difference in driving an automobile at 35 mph and 110 mph. The faster the life cycle of the grass, the more awake and aware you have to be, and the easier it is for you to get in a wreck.

Bermudagrass is a relatively good stocker grass if it is kept in a young and leafy stage. Stocker cattle should not be forced to graze bermudagrass that is over two weeks old! This means the grass must be allowed to grow no taller than three to four inches in height. Buffer grazing is an excellent way to match stocking rate with the variable rate of grass growth.

Bermudagrass cannot be damaged by overgrazing except in the late fall when it should be allowed to go dormant after achieving four to five inches of growth. After frost, this dormant grass can be rationed out with a movable electric fence. After December, the quality of this stockpiled Coastal will require supplementation with cottonseed meal to continue to provide gain to stocker cattle. In very humid climates, consuming this grass by January 1 is recommended to prevent the possibility of mold growth.

Coastal bermudagrass is highly responsive to nitrogen fertilization. Gains in excess of 1000 pounds per acre have been achieved with stocker cattle in a four month grazing season. Care should be exercised with fertilization to not grow more grass than the cattle can eat. As Dr. Bill Oliver at LSU's Hill Farm Experiment Station has said, the primary trick to successfully grazing Coastal bermudagrass is to "Use the forage as fast as it grows and then manage to grow some more forage."

How To Utilize It. Bermudagrass, like most tropical forages, is a good "growing and framing" grass but a poor "finishing" grass. While gains per acre can be exceptional, gains per head are quite low (around a pound a day for a 150 day season-long average) compared to cool-season annuals, and the grazing season is also shorter. However, being a long-lived perennial the cost per pound of gain from nitrated Coastal bermudagrass is frequently half the cost of annual ryegrass. However, the combination of low per head gains and a short grazing season means that a stocker program based upon Coastal alone is financially risky, as not enough total gain is usually produced per animal to offset the rollback in price from calf to yearling.

This does not mean that fertilized Coastal is not an economic stocker grass. What it means is Coastal should always be combined with a subsequent cool-season graze on wheat, oats or annual ryegrass. The highest value of gain in the 1980s was produced from a summer purchase and first quarter sale. Coastal in combination with wheat pasture, for example, would be a good production system to cash in on this high value gain. Remember, always think "Coastal and (a cool-season annual)" and you'll do fine with it.

ALFALFA

University of Kentucky agronomist, Chuck Doughtery, said that stocker grazed alfalfa is "the most profitable legal agricultural enterprise practiced in Kentucky." Doughtery said gains of 1000 lbs an acre were becoming common. "200 bushel corn cannot touch the profitability of grazed alfalfa," he said.

In New Zealand, he said grazed alfalfa pastures had persisted for 20 to 25 years compared to an average of four years under a haying regimen. Alfalfa must be rotationally grazed to persist, but Doughtery said extensive pasture subdivision was not necessary. He said five to six paddocks would suffice and eight seemed ideal. The length of the grazing period should not exceed 12 days during periods of active growth with the possible exception of spring.

He said research in Virginia had indicated that alfalfa could be grazed as long as three to five weeks during the early spring growth period, and that this longer grazing period appeared to help control winter and spring weeds. Grazing periods of only one to two days has shown to allow severe weed encroachment, as the cattle are very selective in their grazing and refuse to eat the weeds. A seven day grazing period followed by a 28 to 35 day recovery appeared to match most farmers' management schedules, he said.

Daryl Clark, a county agent in Muskingum County, Ohio, told me a farmer in his county, Ted Kimpel, produced 2,015 pounds of gain with stocker cattle grazing strip-grazed alfalfa in one year.

Kimpel built a two strand fence of high tensile wire around an eight acre alfalfa field and used polywire interior fences to ration the alfalfa into quarter acre blocks. Kimpel allowed a 30 day rest period following grazing. The eight acre block was stocked with 71 steers weighing 345 to 645 pounds on May fourth and they were grazed until September 6th. The average daily gain was 1.9 lbs per head. The cattle were supplemented with a small amount of home raised corn.

The cost of gain, including a $100 an acre land charge, was only 21 cents per pound. Clark said the stand of alfalfa was thicker and healthier at the end of the grazing season than it had been at the end of the previous year.

Leader-follower Grazing Most Profitable with Alfalfa. Alfalfa grazing is most profitable under a two-phase grazing system, whereby stocker cattle graze the top one third of the plant and cows graze the remainder. Alfalfa should be grazed as close to the surface as possible to trigger regrowth of the plant, but the stems of the plant are much less digestible than the leaves. If stocker cattle are forced to graze the whole plant, their average daily gain will be severely depressed. However, if the stockers are only forced to graze the top one third of the plant, average daily gains can exceed three pounds per day! Such exceptional gains require a very low stocking rate, and a two pound per day gain can be achieved with a 70 percent plant utilization and a much higher stocking rate.

Two Sets Of Stocker Cattle. The carrying capacity of alfalfa pastures in the summer/fall phase will be 25 to 50% below the spring period, and is why a declining stocking rate must be planned for. For example, spring alfalfa pastures could be used to "finish" a set of steers that have overwintered on stockpiled fescue, and at the same time "start" the summer set of replacement steers. During rapid growth phases, some alfalfa may have to be routed to hay or silage to prevent the pasture from becoming overly mature.

ALFALFA and FESCUE

Alfalfa grazed in sequence with separate pastures of stock-piled tall fescue could provide a 270 day grazing season with an overall average daily gain of two pounds per day in Kentucky, said Chuck Doughtery. Soil erosion and treading damage are reduced if alfalfa is grown in conjunction with tall fescue, orchardgrass, or bromegrass. Sacrifice paddocks should be planned for use during excessively wet weather to prevent treading damage. No supplemental feeding should ever be done on the alfalfa pastures due to the potential for treading damage. Feed only on the sacrifice area. Doughtery said bloat was not a problem as long as Rumensin was fed to the cattle. A mineral mix high in calcium and sodium is necessary due to the deficiencies of both of these minerals in alfalfa.

DOCK Prevents Bloat

Research at Massey University in New Zealand has confirmed many graziers' observations that the common cool-season pasture weed, dock, does prevent bloat in cattle grazing alfalfa, red and white clover and, in fact, when consumed in small amounts actually enhances the nutritive value of these forages.

Dock and non-bloating legumes such as lotus, sainfoin, Birdsfoot trefoil and crown vetch have been found to contain a substance known as condensed tannins (CT). CT not only prevents bloat when consumed in small quantities but helps to precipitate protein from consumed forages. Gains in productivity of 10 to 15 percent are thought to be possible by including CT carrying forages in a pasture mix.

The Massey researchers hope to eventually be able to genetically alter forage legumes to include CT, which will not only render them non-bloating but allow increased animal gains and milk production.

BRASSICAS

Brassicas include turnips, rape, Tyfon, swedes and kale. They can be used to increase mid-summer forage availability but are most useful in the late fall and early winter period to bridge the quality gap between summer perennial pastures and cool-season annuals.

Advantages:

☆ High protein and high digestibility.

☆ Retains nutritive value at advancing stages of maturity. Lends itself well to stockpiling for later use.

☆ Relatively acceptable to livestock.

☆ Tolerates low temperatures.

Disadvantages:

☆ Low in fiber, which causes a tendency toward diarrhea.

☆ High in calcium and potassium; but borderline in copper, zinc and magnesium.

☆ Needs warm temperatures for planting (similar to corn).

☆ Not competitive with other plants at establishment. Conventional or chemical tillage necessary.

Recommended Management:

☆ Summer grazing should be light. Top grazing only.

☆ Graze small areas at a time using management-intensive grazing.

☆ Combine with cool-season annuals to prevent erosion and provide needed fiber.

☆ After maturity, fungal diseases and rot may deteriorate brassicas.
☆ Brassicas should be planted in the same area for only two years
due to potential for disease buildup.

WARM-SEASON ANNUALS

Most disappointments with yearling summer gain performance with warm-season annuals is due to the fact that the pasture
growth gets ahead of the cattle, and the forage becomes stemmy
and undigestible. Graziers must monitor the growth of their forage
and move the cattle as fast as the growth demands. If the growth
starts to get away from you, drop a couple of paddocks and shred
them or make hay from them.

With irrigation, four or five paddocks probably will be all
you will need, as there will be no need for forage rationing. Keep in
mind, a large number of paddocks hinders rather then helps when
forage growth is rapid.

With summer forages, Brahman cross cattle will produce
gains of a half pound a day or so better than English or exotic
crosses. Also, very few warm-season forages can produce good per
head gains on calves, and so are best grazed with yearling cattle.

Now, I want to briefly review some summer annual forages
and their use with stocker cattle. All of these forages require
leaving a high residual to produce high average daily gains. Stocking rates should be experimented with to find the best one for you,
although your seed supplier or local livestock extension agent may
have a good rule of thumb for your area.

TIFLEAF MILLET: This African grain can produce good
per head and per acre gains with yearling cattle in hot climates. It
does not produce good per head gains with calves. It also does not
do well in climates where the daytime temperature is consistently
below 90 degrees F (30 degrees C). If it is too cool to grow cotton
in your area, it is probably too cool to consider Tifleaf millet. With
irrigation and supplemental nitrogen, gains per acre can be exceptional.

COWPEAS: Known as black-eyed peas in the South, this
warm-season annual legume is cheap to grow, very reliable and will

produce good per head gains on yearling cattle. It does not produce good gains with calves nor does it produce very high gains per acre. Cattle must be forced to eat cowpeas initially by confining them to a small paddock of peas. A pea-wise older bellcow or lead steer can quickly show new cattle the peas are good to graze. Cowpeas grow best in hot sandy, cotton-type soils.

HAYBEANS: Also known as fine-stemmed soybeans, haybeans are also cheap to grow, are reliable, and offer a slightly higher average daily gain than cowpeas but require a better soil fertility and pH than cowpeas. Haybeans can be grown wherever soybeans are grown. Gains per acre are relatively low but so are costs.

SORGHUM SUDANGRASS: This annual can be grown successfully in cooler climates than dwarf millet. Grazing management is crucial due to the plant's extremely rapid growth. The grass produces good gains on yearling cattle and gains per acre under irrigation and supplemental nitrogen can be exceptional. Possible problems with nitrate toxicity can be avoided by leaving a high residual of eight to twelve inches.

CRABGRASS: This Scandinavian "weed" is the only warm-season annual to my knowledge that will produce high average daily gains with calves. Stocking rate and grazing management is similar to that for winter annuals. Crabgrass' poor reputation for reliability is caused primarily by a lack of spring soil disturbance on the part of the grazier. A light spring disking and harrowing is necessary for the best results with crabgrass. Crabgrass residue is toxic to new seedlings and must be removed with surface tillage to produce a new crop. Ryegrass and cereal rye residues are also toxic to crabgrass and will greatly slow down the onset of the crabgrass if not removed with tillage. Crabgrass is a very reliable plant when spring surface tillage is practiced and moisture is adequate. Crabgrass will grow virtually everywhere in the United States and Canada. There is a great deal of yield difference in crabgrass varieties. A very high yielding variety is "Red River."

Dual Purpose Lowers Cost of Gain. Due to their short growing season, warm-season annuals tend to be a relatively high cost of gain forage compared to cool-season annuals or perennials. However, when used both as a smother crop and as grazing crop for gaining animals, they can be very economical.

SMOTHER FORAGE CROPS

Deny a plant access to the sun and it dies. This is the basic idea behind a smother forage crop. In Argentina, smother crops are the primary way weeds are controlled in both pasture and crop fields. The primary plant the Argentineans target with smother forage crops is common bermudagrass — a plant impossible to control with tillage. In North America, the primary target for smother crops is fungus-infected fescue.

Don Ball, extension forage agronomist with Auburn University, said a smother crop of sorghum sudangrass or pearl millet is an excellent way to prevent "escapes" from a spring tillage/herbicide, fescue removal regimen. Don said the ability of these warm-season plants to go from seed to "head high" in just a few weeks insured that they would shade out new fescue seedlings. He said a key aspect of this program was to kill the fescue before it made a seedhead in the year it was being removed.

Due to the high, but variable, stocking rate these annuals require and the extra management the subsequent new grass stands demand, it is probably best if no more than 5 to 10 percent of your pastureland is undergoing a smother crop renovation in any one year. A staggered planting date will help to put a "cut" in the forage than can be maintained with management-intensive grazing. This "cut" can be maintained with as few as 5 paddocks.

SORGHUM SUDAN HYBRIDS AND SUDANGRASS. This North African leafy annual will grow four to eight feet in height. It is very drought tolerant but cannot stand highly acid soils. In Alabama, Don said it does best on heavy clay, blackland soils. Seed are drilled at 20 to 25 lbs per acre or broadcast at 30 to 35 lbs in May or June. It normally grows until September. Some varieties are very late maturing and do not normally make a seedhead.

PEARL MILLET. This annual is from North Central Africa and has a rapid, erect growth similar to Sudan but is much better adapted to sandy, acid soils than sudangrass. A real heat lover, it has a slightly shorter growing season than sudan. Seed are drilled at 12 to 15 lbs per acre or broadcast at 25 to 30 lbs per acre. Both sudangrass and Pearl millet are highly responsive to nitrogen.

Grazing Management. Both sudangrass and Pearl millet are very high in quality if grazing is initiated at the 24- to 40-inch high stage. Average daily gains on yearling steers are frequently near 2 lbs a day and milk production is higher than on warm-season perennials. These annuals make excellent "popping paddocks" for stocker cattle coming off fungus-infected fescue.

At LSU's Red River Station, 750 lb yearling steers coming off of ryegrass made summer gains in excess of 2.5 lbs per day on rotationally grazed sorghum-sudan. These heavy steers were continuously stocked at two per acre with no on-off grazing. Such fleshy cattle traditionally gain poorly on warm-season perennials after mid-July.

A minimal residual of at least 8 inches should be maintained at all times both for the health of the annual and to maintain its shade weed-control function. The recommended stocking rate is unspecifically described as "high." Due to its rapid and erratic growth rate (due to rainfall variation) both of these annuals are best used under a combination rotational/on-off/put-and-take grazing program in combination with other perennial forages. An on-off stocking creates the necessary high — but variable — stocking rate needed to keep the annuals in their "quality window" of 15 to 30 inches. Without a very high stocking rate these annuals are easy to get away from the cattle and become overly mature. This will collapse animal gains.

Being of African origin, both of these plants need heat to grow and therefore grow best south of the Ohio River. This is particularly true of Pearl millet. Sorghum sudan can be grown virtually anywhere corn can be grown. Many graziers find these warm-season annuals to be highly tolerant of residual herbicides in the soil and use them as a "cleansing" crop prior to planting more sensitive cool-season grasses.

Don Ball said in Alabama these annual crops are traditionally cut for hay in September to remove the plant residue that would interfere with fall cool-season grass planting. Due to their thick stems both annuals are difficult to cure for hay and are better when removed as silage.

Nitrate Poisoning. Animals grazing heavily nitrated sorghum sudan or Pearl millet under droughty conditions are susceptible to nitrate poisoning. Animal symptoms are labored breathing, muscle tremors and a staggering gait after which the animal gasps for breath and dies quickly. Prompt medication with a 4 percent solution of methylene blue supplied intravenously using 100 cc per 1000 lbs of body weight will prevent death.

Nitrate poisoning normally occurs when very hungry animals break through a fence and gorge themselves on freshly nitrated, droughty forage under 15 inches in height. There are few problems from animals grazed at high plant residuals. Hay and silage cut from nitrate-toxic forages **does not** become less poisonous with time. Probably more cattle are killed with high nitrate hay and silage than from grazing. This is particularly a problem in large roundbales and silage stacks where "hot spots" of high nitrate forages can hide from forage testing. Corn, wheat, oats and the weeds pigweed, smartweed, ragweed, lambsquarter, goldenrod, nightshade, bindweed, Canada thistle and stinging nettle are all potential sources of nitrate poisoning as well.

Prussic Acid Poisoning. Pearl millet does not produce prussic acid but sorghum sudangrass does. Prussic acid is most likely to reach dangerous levels immediately following a frost. However, this can also occur after the use of herbicides. Symptoms include excessive salivation, rapid breathing and muscle spasms. Death can occur within 10 to 15 minutes after consuming the plant.

Prussic acid and nitrate poisoning **are not** the same. Unlike nitrates, prussic acid deteriorates with time. Hay and silage older than 3 weeks are safe to feed. Standing plants are safe after one week following frost. However, be observant. Low land areas frost first. Sloping land and hilltops can cause problems with later and more severe frosts. Johnsongrass and wild cherry are other prussic acid problem plants.

Direct Grazing GRAIN CROPS

In Australia, the grazing of grain sorghum with beef cattle is very common. It is used primarily as a "finishing forage" during the late summer/early fall period when good gains are difficult to produce in hot weather regions from perennial forages.

In the tropical, summer/wet regions of northern Australia, grain sorghum crops seldom field-dry enough for machine harvest. Grazing the grain sorghum bypasses both the expensive grain drying and the need for a "heavy metal" combine. Since the crops' nutrients are largely recycled in the manure, fertilizer costs are dramatically lowered.

Grain sorghum is much more effectively utilized by yearling-aged cattle than by calves. It is most cost-effective when used to fill a specific gap in a lower-cost perennial based forage system.

Grain sorghum recovers quickly from grazing if only 50 percent of the plant is taken per rotation. This lax grazing also helps the animals maintain a high rate of gain (two pounds per head per day with Brahman-cross yearlings), and negates any potential problems from prussic acid poisoning. The use of animals with 50 percent Brahman breeding has been shown to improve average daily gain by at least one half pound per day in hot climates on grain sorghum.

FINISHING CATTLE ON GRAZED GRAIN SORGHUM

For finishing cattle, the rotation should be slowed so the grain sorghum can make a grain head. This grain-head should be grazed while it is in the greenhead stage and not allowed to mature into the hard grain stage.

With a slow rotation, two grazings in the green grain-head stage are possible, and one or two in the green leaf stage, depending upon the length of the growing season.

Frosted grain sorghum stubble makes a good dry-cow wintering forage and is widely used for this in Australia. The Australians consider grain sorghum to be a primary forage crop in summer hot/dry or hot/wet climates. It is drought tolerant, has the ability to rapidly regrow following grazing, produces multiple seed heads, has a long growing season, and has dual usage as a summer finishing crop and as a standing hay crop (foggage) in the winter.

CORN

Across the Tasman Sea, the New Zealanders with a more temperate climate use corn (maize) as a supplement for their dairy cows during the mid-summer slump in perennial ryegrass production. Corn can be grown and utilized as a grazing crop in climates where the growing season is too short for a regular grain crop.

Corn (maize) is a grass. It is from the same family of warm-season grass plants as Big bluestem, Johnsongrass and Eastern gamagrass. When we start to graze it rather than grow it for grain, the same management rules of thumb that apply to all other grasses also apply to corn.

As with all other grasses, only immature green leaves produce high rates of gain or milk production. For high animal performance with corn, the plant must be immature. It must be green. The animals should not be forced to graze the indigestible woody stalk portion of the plant. The grain portion of the plant is relatively unimportant with grazing because it makes up such a small percentage of the total plant.

Corn as a forage has two major attributes. One it is the highest producer of dry matter per acre of any grass, and two, it has a relatively long, quality-forage, time window of 60 days. This means a corn crop can be slowly rationed out with strip grazing without losing quality to plant maturity. Whereas, most grasses and legumes are high in protein and low in carbohydrate, corn is just the opposite. This makes it an ideal summer supplement to cool-season pasture for dairy cows and yearling beef cattle. Any corn that is not utilized in the green season can be used later as a standing winter feed crop for dry cows and yearling cattle being "roughed through" to spring pasture.

It will take 10 days to two weeks for the animals' rumens to become fully adapted to the high carbohydrate forage so always give them access back to pasture. Let them decide how long they want to graze the corn. Again, beef gain and milk production from grazed corn comes primarily from immature green leaves. Initial turn-in should be at the very first sign of tassel formation. In most cases, this will be at about shoulder height. Corn at this stage will be at about 15 percent dry matter and will gradually increase in dry matter for the next 60 days.

The corn should be tightly rationed. Any sign of bite and walk and end to end milling, indicates a too-large break. The animals should graze across the break face as if at a feedbunk. It will take some time to match your break size to the number of animals. Hang loose and play with it.

Use a tractor or ATV to knock down a swath of corn to run your one-wire temporary electric fence through the corn to make the break. Make sure the fence does not touch any of the standing corn as this will energize the corn and give your animals a nasty surprise. Remember, we want the animal to be able to see and know where the fence is so as to not psychologically restrict his grazing enthusiasm. With an adequate break, most animals will be completely full within three hours. Allow them access back to permanent pasture.

Watch the animal's manure. It should be of sheet-cake consistency for high animal performance. Lumpy, hard manure indicates the animals are grazing too much stalk. Remember, it is green leaves that create gain and milk.

Corn grown for grazing and silage is at the opposite end of the plant selection spectrum than corn grown for maximum corn yield, said Cherry Fork, Ohio, corn breeder, Dick Baldridge. "Corn that is grown for grain is bred to have a high-lignin stalk to prevent lodging. This high lignin makes a grain-type selection exactly backward for what we are looking for in a good forage plant." He said a traditional corn farmer would not consider the bushy look of a good forage corn to be very pretty but might find the net returns per acre pretty attractive. He said the net return per acre from grazing corn can be four to five times higher than with grain production.

Baldridge has been making his forage corn plant selections based upon low lignin, high numbers of soft leaves and high plant

protein content (10 to 13 percent protein). He has little to no
concern about the plant's grain yield. "I have found little correlation
between grain yield and animal performance with corn silage. The
correlation is with the amount of leaf, lignin and the maturity of the
plant. Or just as it would be with any other forage plant."

Use high plant population. Just as with other grasses, we
want a very high plant population per acre. We don't want to waste
any sunlight and we want to use the shade produced by tight spac-
ing to prevent grass and weed competition. Corn for grazing should
be drilled in seven inch rows or planted with a spin seeder at 40,000
to 50,000 kernels per acre on a prepared seedbed. Your mental
picture of a field of corn grown for grazing should be one of a five
to six foot tall dense pasture rather than a widely-spaced corn field
with lots of bare ground showing.

A corn grazing pasture is not a low-input crop. Figure at
least $120 to $150 input cost per acre of corn grown for grazing.
(See sample budget.) Because of the high plant population and
short growing season, it requires a lot of initial nitrogen. Recommended minimum rates are 400 lbs of ammonium nitrate per acre or
600 lbs of ammonium sulfate per acre. (150 lbs of actual N.) Urea
is not recommended for grazing corn.

Budget for No-Til Forage Corn

Nitrogen	$52
Seed	$36
Crop Oil	$5
Machinery	$20
Lime, P&K	$20
	$133

For conventional tillage, add an additional $25.
Machinery cost is for custom hire.

Yellow-green, low nitrogen corn will be very low in protein
and can make animals scour if forced to graze it. Nitrogen toxicity
is no problem with Baldridge Grazing Maize if only one initial
nitrogen application is used as recommended. As with all other

pasture inputs, grazed corn requires an increase in stocking rate to be cost effective.

"Corn should never be used to replace low cost perennial pasture. It should primarily be used as a supplement to it," Baldridge said. "In our computer simulations of the grazing season, we were able to increase our spring stocking rate 20 to 40 percent by having corn to supplement the pasture during the summer slump. This will allow graziers to utilize more of their high quality spring pasture."

With stocker cattle, corn can be used to both keep cattle gaining well in the summer and to carry them out of the low priced "summer slump" market runs that occur whenever pasture runs short in July. Ideally, heavy stocker cattle should be sold directly off corn, so that more of the autumn pasture can be used for stockpiling and winter use with lighter cattle.

Baldridge said an acre of forage corn with 15,000 lbs of dry matter would feed 8.3, 800-lb steers for 60 days if it was their only feed source. If the grazing was limited to only three hours each morning, the same acre could supplement more than twice that many. The protein content of Baldridge's forage corn is adequate for 750 lb plus, early maturing cattle, but may not be for late-maturing breeds such as Limousin and Charolais at the same weight.

In the no-til corn-into-permanent-pastures I saw in the Midwest, there was enough volunteer white clover growing under the corn to provide more than enough protein supplement for yearling beef cattle. Keep in mind, a three-hour morning corn graze will greatly dilute the effect of summer fescue toxicity on stocker cattle.

For dairy cows, one acre of forage corn per 10 Holsteins would be adequate for a morning-only grazing program. The protein content is not high enough for a corn-only diet and access should be given to leguminous-pasture.

Baldridge recommends that corn be used for the morning graze and permanent pasture for the evening graze. "We don't want the cows blundering into an electric fence at night," he said.

For a longer than 60 day grazing season, the corn should be stagger planted at three to four week intervals. Baldridge recommends a May 1, June 1 and July 1 planting regimen in the Midwest.

In the Deep South where drought-tolerant tropical corn is best, Baldridge recommends a June 1 planting following annual ryegrass and a July 1, August 1 subsequent planting. He said such a staggered planting should largely eliminate the autumn "hole" in quality forage production that plagues annual ryegrass-based dairy and stocker graziers.

Baldridge has developed specialist forage corns for both the temperate and sub-tropical regions of the United States. These are marketed as "Baldridge Grazing Maize." He is currently working on a very early maturing variety for Canada and New England and a high leaf, low-lignin forage rye that will provide a winter-double crop with grazed corn.

Tips for No-til Seeding of Corn into Pasture. Plant corn with a no-til drill on seven inch spacings. Burn down surface vegetation with crop oil and 28 percent liquid N at the rate of 150 lbs of actual N per acre. This will not kill the pasture but produces an artificial "frost" that will allow the corn to get ahead of the pasture regrowth. If liquid N is not available broadcast ammonium nitrate at the rate of 400 lbs per acre and then immediately spray pasture with crop oil.

Planting corn into pasture will not hurt cool-season pastures and legumes but will kill warm-season pastures due to shading. Forage corn is an excellent way to remove bermudagrass and other low-growing warm-season weeds. Of course as most corn farmers know, due to their genetic similarities a corn crop is an excellent way to establish or thicken a crop of high quality Johnsongrass.

Innovative Uses for Forage Corn. *LAMBS AND STEERS TOGETHER*. A two-tier grazing system can be used whereby lambs are first allowed to graze the lower leaves of the corn and yearling cattle are subsequently allowed to graze the upper leaves. These lower leaves will fall off due to shading if not grazed. Waste not want not. The two-tier grazing is best in no-til situations where there is an understory of white clover and cool-season grass as the

corn leaves alone will not have enough protein for the lambs.
FINISHING STEERS. Early maturing cattle can be finished
on grazed corn if the animals are either very heavy when first
turned into the corn or a stagger-planting regimen is used to give a
longer grazing season. As long as the animals are still growing, a
time-limited graze is best. However, for fully grown animals (850
lbs for an Angus steer) a full graze on just corn will finish them
faster. An opportunity in these high corn price days is to buy feeder
weight cattle (700 plus) in late May or July and put a cheap 200 lbs
on with grazed corn before taking them into the feedlot for a fast
finish. There is much less rollback to eat with these heavy steers
than with lighter more traditional weight steers.

OATS and BARLEY

Oats and barley are two grain crops that were once widely
grown to be direct-harvested with (primarily dairy) cattle. These
two crops are best utilized when interplanted with late maturing red
clover. This grain-legume combination was said to really boost the
milk production of a fall-freshened cow nearing the end of the
lactation.

Headed-Out OATS and WHEAT

Stocker graziers in central Texas have long been cursed by
cattle feeders for "finishing" their cattle by allowing them to graze
headed-out oat crops prior to selling them as heavy feeders. While
this practice is not recommended as a way to build a reputation as a
feeder cattle supplier, it should definitely be considered for those
planning to own their cattle all the way to slaughter.

Wheat growers should also keep in mind that standing head-
sprouted wheat is an excellent crop for direct-grazing with yearlings and could be a good way to salvage a rain damaged crop.
With yearling cattle, gains are best when the plant and grain are
harvested in the immature green stage or after the mature grains
have head-sprouted.

The rest of the world thinks we North Americans are pretty
crazy to be hauling feed to drylotted cattle during times of the year

when there is plenty of green feed that can be direct-harvested. Perhaps the day will come when we think it is pretty silly too.

Pasture Profits
Look for these signs on a pasture walk:
* No clover or clover with small leaves indicates a need for calcium.
* Low calcium soils will not grow stocker cattle well.
* Dandelions are an indicator of low-soil calcium.
* Clover growing better in dung pats indicates low phosphorus.
* Clover growing better in urine patches indicates low potassium
* Grass growing better in urine patches indicates low nitrogen or sodium.
* A thatch on the soil surface indicates the lack of earthworms.
* Liming will double earthworm numbers in three months.
* A healthy earthworm is slimy and clean. If soil clings to them it indicates a need for lime.

Chapter 15

Manipulating Pasture Quality
With Grazing

"The mind of the intelligent gains knowledge; and the ear of the wise seeks knowledge." Proverbs 18:15

In New Zealand, 60% of the total annual grass growth occurs in just three months of the year. However, with a topography similar to Colorado, much of New Zealand's grazing land is too steep to be harvested with machinery. How do they deal with this huge annual forage imbalance? With beef cows.

"The secret to managing spring pasture surpluses without machinery is beef cows," explained Dr. David McCall of the Whatawhata Research Center. Whatawhata is a "whole farm system" beef and sheep grassland research center located in the steeply rolling hills east of Hamilton. While admitting that beef cows return less per acre than sheep or (stocker) finishing programs, he believes sheep and stocker programs can make more money with them than without them. "The beef cow is the critical pasture tool in non-machinery areas," McCall said. "The two major management components we have to work with are the time of calving and the use of cow body-fat reserves."

McCall pointed out that the beef cow's feed requirements

and intake nearly triple once she calves. This is a real benefit in helping to manage spring pasture surpluses. By timing the calving close to the onset of the spring growth, the pasture naturally provides this huge increase in feed supply. Another factor is that the cow's natural ability to get in-calf improves the closer she calves to the longest day of the year. He said for the best use of spring pasture the calving season should be no longer than eight weeks in length.

"Calving timed to coincide with the surge in late spring pasture growth, improves herd reproductive performance, thus requiring fewer heifers to be brought into the herd. It also allows a higher stocking rate of cows to be carried through the winter. This ensures more of the spring surplus pasture growth is eaten, which benefits summer-autumn pasture quality. The economic balance favors producing more calves from later calving," he said.

"All low-cost production methods are very seasonal in nature. You can't have low-cost and year-round production. In pastoral systems, animals must fit the pasture growth curve to be profitable."

McCall said that the second benefit of beef cows was the ability to use a cow's body-fat reserves to match animal and pasture needs. He said to maintain pasture quality acceptable to stocker cattle, paddocks that cannot be grazed before they reach 12 cm (5 inches) in height should be dropped from the rotation and grazed later with cows as "standing hay" in non-machinery areas.

To facilitate this, a major goal with a spring calving program should be to allow your cows to get as fat as possible by mid summer. This reserve of body-fat will allow the beef cow to clean up rank growth in the paddocks dropped from the spring rotation while still producing acceptable calf growth in her suckling calf.

"Unweaned five- to seven-month-old calves grazing low quality pasture grow faster by 0.2 kg/day (half a pound) than calves weaned at five months onto hay aftermath," he said. Up to 35 to 40 kg (77 - 88 lbs) can be stripped off the cows' backs to keep the calves growing at an acceptable rate. Once the surplus is removed the calves should be weaned and the cows dropped to maintenance.

Conversely, if spring growth rates are slowed by drought and no surplus pasture is created, the calves can be short-weaned much earlier and the cows dropped to maintenance to save scarce summer grass. This ability to dramatically and quickly alter feed demand makes the beef cow an excellent "buffer" animal for other higher return enterprises. McCall said the great conflict all graziers face is between what the market wants and what the pasture needs. "Hill country farmers want to go completely to finishing beef or lambs because that appears to be the greatest return per acre. But you can't maintain your pasture productivity long term with such simple single enterprise systems."

Sheep and Cattle Together Offer Best Harvest System

For example, "cattle only" systems are 10 to 25 percent less productive than those that combine cattle and sheep. He said cattle only systems cause pastures to lose density and become weed infested. Conversely, "sheep only" systems cannot control the spring lush, and ewe and lambs' performance is much better with cattle than without them. Grazing sheep and cattle on separate blocks in alternate years not only provides both good pasture weed control but also helps lower parasitism in both species.

At Whatawhata, a 60/40 mix of sheep and cattle on a weight basis appears to be the most profitable combination. Like the beef cattle, lamb births are closely timed to the start of the spring lush, and they are weaned at 10 weeks of age and sold to specialist lamb finishers in lowland areas. Similarly, growing beef animals (stockers) are kept tightly reigned in during the winter and then are allowed to compensate on the surplus spring pasture. Spring growth rates in excess of four pounds a day are achieved in the spring and early summer.

Fertile Pastures Have a More Even Growth Curve

Keeping pasture fertility high helps to smooth out the pasture growth curve. Poor fertility pastures that have not been regularly fertilized with lime and phosphate have a much steeper spring growth "spike" than those that receive the "little and often" method of pasture fertilization. Nitrogen is not considered a fertilizer in New Zealand but is categorized as a "feed supplement." Nitrogen is used sparingly to fill "holes" in the pasture growth curve rather than on a regular basis.

In the United States it is estimated that 70 percent of the annual cash cost of a beef cow is her winter feed bill. The second major use of cow body-fat reserves is to use them to reduce winter feed requirements, explained Dr. David McCall of the Whatawhata Research Center. This not only saves feed but allows the cow to be used for weed control and pasture improvement. "We've found that the body condition of a late calving cow can be reduced in winter without a great effect on calf weaning weight," he said.

Dry Cows Provide Thistle Control

By combining the dry cows and dry ewes into a large mob at a very high stock density, competition forces the cows to eat less preferred plants and weeds. For example, such tight, non-selective wintergrazing helps provide natural thistle control.

Calf Grafting Systems

While at Whatawhata I saw an experiment designed to increase the cost effectiveness of beef cows by grafting an additional calf onto the cow. The North Island of New Zealand is rapidly becoming dairy country and more and more of New Zealand's beef cows are the genetic result of using easy-calving, beef bulls on first calf dairy heifers. Of these crosses, the Friesian/ Hereford is most popular as its white face clearly signals its beef breeding component whereas the Friesian/Angus cross is often marked similar to a straight Friesian. Most of these dairy-cross cows are bred to terminal sire beef bulls and all replacements are purchased.

The Friesian component of these cows in combination with lush spring pasture produces more milk than a single calf can utilize. As a result, Whatawhata has been experimenting with grafting another calf (usually a male dairy calf) onto the cow to utilize the excess milk. Since New Zealand's dairy breeding is highly seasonal as well, there are around five million newborn male dairy calves to pick from in the spring. Most of these are currently killed at four days of age as vealer calves.

The dairy calf is chained to the cow's natural calf with a short length of chain shortly after birth. Peter Moore of the Station said they have had a 96 percent "take" using this method. By nursing two calves, each cow produces an additional 300 lbs of beef that is essentially at no additional feed cost. The additional calf nursing does force the cow's intake up, but because calving is timed to the spring lush this helps control the spring pasture surplus. The calves are weaned at the end of the spring lush at three to five months of age and sold to lowland beef finishers. Moore said they have seen no significant difference in the rebreeding rate. "It appears it may actually be better than our cows nursing a single calf."

In conclusion, Whatawhata feels that beef cows have a definite place on all humid, temperate climate enterprises seeking to eliminate machinery costs. While they should probably never be the centerpiece enterprise in a humid, temperate climate due to their low return per acre, they are very complementary to other higher value ruminant enterprises in creating and maintaining quality pasture.

Set Stock or Rotational Graze during Spring?

Dr. McCall said that rotational grazing is primarily a rationing tool best used during slow growth or no growth times. However, a fast rotational graze in spring will maintain pasture quality better than set stocking.

Animal intake is most easily maximized with set stocking. However, when animals are given access to large areas of pasture, dead material at the base of the sward will accumulate and produce very poor quality grazing later in the summer.

A fast rotation with no more than 21 to 25 days between grazings is the best compromise between maximizing intake and controlling future quality. Also, surplus pasture accumulation can be more easily recognized when it is concentrated on individual paddocks rather than spread across the farm.

Set Stocking For Spring Pasture

Vaughan Jones pointed out that most New Zealand graziers set stock during calving and lambing, which are carefully timed to coincide with the start of the spring lush. "Just because you've got a hammer doesn't mean you have to use it on every job. There are times to intensively graze and there are times to set stock. When your pastures are growing very fast it is best to set stock and allow the animals to top graze."

He said pasture fertilization must be timed so as to not increase unwanted pasture surplus. The best time to fertilize is in the fall to build winter pasture and in the late spring to build pasture for the summer slump. Nitrogen should always be used in small applications.

He warned the heavy use of fast-release super phosphate and triple superphosphate made both the pasture and the grazing animals sick. "Pasture is not like a rowcrop. We don't want to jump it all up at once. We need a nice slow release fertilizer source like naturally reactive North Carolina phosphate," he said.

Feeding Salt to Your Pastures Can Make Them Taste Good

He said pastures topdressed with hog or chicken manure need salt (sodium) as well to balance potassium buildup that naturally occurs. Also, topdressing pastures with small amounts of salt has shown good results in New Zealand by increasing animal intake. Apparently, the salt makes the grass taste better to the animals.

Another major, often overlooked fertilizer was lime. He said lime had been talked about only in terms of pH, but pastures need calcium, many desperately so, even if they have a high pH. (In many areas gypsum is available much cheaper than lime, and will provide soil calcium and sulfur without raising the pH.).

Liming pastures also promotes earthworm growth. Unfortunately, cheap fertilizer elements like salt, lime and natural rock phosphate are seldom promoted because of the lack of profit to the major fertilizer companies.

Vaughan said as far as he was concerned a soil test was only useful as far as the pH was concerned. He said testing your soil once a decade should be more than adequate under grazing conditions. What was important was the forage testing grass since that was what the animal was eating.

Vaughan said that as your stocking rate increased, so did the importance of your mineral supplementation program. He said most of his clients have gone to water soluble minerals fed to the cattle through the pasture water supply based upon the forage test.

His dairy clients are seeing tremendous increases in milk production through the balancing of minor trace elements in both the pasture via fertilization and in the animal by supplementing minerals through the water supply. Selenium, for example, has been found to greatly reduce the negative effect of the endophytes in perennial ryegrass and fescue.

Another way to minimize endophyte problems is to not graze the last three to four inches of fescue and perennial ryegrass. He said there is very little endophyte in the top one-half of the plant.

Keep in mind when stockpiling pasture for wintergrazing, after 8 to 10 inches in height, every inch you allow to grow on top of the plant is subtracted from the bottom of the plant as the leaves die and drop off from lack of sunlight.

These dead leaves acidify the soil, kill clover seedlings and promote unhealthy fungus and mold growth. To prevent this, he said to topgraze the stockpiling pastures with a few animals and keep them from getting too tall and thereby preventing sunlight from getting to the base of the plant.

High Stock Density Grazing Improves Pasture Plant Diversity

The Ravenscroft family's 58,000 acre ranch, 3-Bar Cattle Co., lies deep in the heart of the Sand Hills of western Nebraska. These hills are made up of bright white sand that reminded me of the snow-colored sand dunes of the Florida Gulf Coast. The only thing that keeps the Nebraska sand hills from blowing and moving like a desert dune is their covering of grass.

Maintaining this grass cover is a primary management goal

in the region. It has been widely thought that the best way to do this was with a low stock density per acre. However, the Ravenscrofts have found that just the opposite — a high stock density per acre — provides both the best grass cover and the highest animal performance thanks to the re-emergence of warm-season prairie grasses in the predominantly cool-season annual pasture.

First some definitions. Stocking rate is the long-term stocking rate for the entire grazing area. For example, one steer per four acres for six months is a typical stocking rate. Stock density is that same stocking rate confined to a much smaller area for a very short amount of time. For example, 1000 head on 20 acres for half a day.

Research at the Noble Foundation in Oklahoma has found that high stock densities (minimum of 40,000 lbs per acre) trigger the growth of warm-season native prairie grasses such as Big bluestem, Little bluestem, Switchgrass and Indiangrass. These grasses not only provide very high quality spring to mid-summer grazing but stand upright through the snow for use as standing hay. Why these grasses appear without planting when high stock density grazing is initiated is not known. It is supposed it is because these grasses evolved in concert with high stock density bison grazing.

High Stock Density Grazing Mimics Nature

In pre-historic days, wherever lightning had created a small burn bison concentrated in large numbers to lick the ashes for minerals. They later returned to these small areas because the grass would be less mature than the surrounding prairie. What was often overlooked was that most of these fires were relatively small in area and occurred throughout the year. Later, native Americans mimicked nature's fire ecology and used small fires to attract bison for hunting. While the role of fire in the creation of a prairie ecosystem has been widely documented, only recently has the importance of the companion phenomenon of high stock density been recognized.

Under modern management, stock density is primarily raised by paddock subdivision and time controlled grazing. With a constant size herd, the smaller the paddocks become, the higher the stock density and the shorter the grazing time. The big question in semi-arid regions where stocking rates are necessarily low is, how can one afford to fence the ranch down into paddocks small enough to create the necessary high stock densities? The answer the Ravenscrofts have found, is with the use of low-cost, temporary electric fence.

John and Jack Ravenscroft currently graze 5400 head of beef cattle on their ranch, 900 of which are brood cows and the remainder calves and yearlings. This is twice as many cattle as they grazed ten years ago before they began to subdivide their ranch. "We found that running twice as many cattle is not only cheaper because your overhead costs are cut in half, but actually easier. We now have only one full-time hand where we used to have three to handle half the number of cattle," John Ravenscroft explained.

Most of the increase in stock numbers has been by the addition of purchased stocker cattle. Target sale weight of both home-raised and purchased cattle is 850 lbs.

John and his brother started subdividing the ranch in 1985. Permanent subdivisions number120 large paddocks. These are then cross-fenced using plastic step-in posts and roll-up electric wire to create the desired high stock densities. Jack has modified an old flat-bed pickup into a fencing truck. On one side is a large fire hose wheel, which is used to roll-up and let-out the electric fence wire. Jack and his son, Andrew, have perfected the two-man technique that can build or take down fence as fast as you can walk.

The ranch consists of both sand hills and naturally sub-irrigated, cool-season perennial grass meadows. It is on these lush meadows where most of the stocker grazing is concentrated and a season-long stocking rate of one steer per acre is used. Average daily gain for a May to October graze on these meadows is in excess of 2 lbs per day.

The yearling stocking rate in the sand hills is 8 to 9 acres per steer. Most of these paddocks are only grazed once during the growing season with the yearlings and twice with the cows. Thanks to the volunteering in of native warm-season grasses, season-long (May - October) average daily gains have risen to 1.7 lbs per day.

The stocker cattle are managed so as to only graze the tops of the plants and leave the rest for the follower cows. Prior to the use of high stock density grazing, the range was dominated by cool-season annuals that produced very poor quality grazing after mid-June.

The ranch windmills were totally inadequate, and two diesel-powered portable pumps have been built to boost stockwater flow during the May to October stocker season. These portable pumps move with the stocker cattle from watering point to watering point. During the winter months, the windmills alone provide adequate water for the dry cows.

Thanks to the shifting of the calving season from March to May/June, it has not been necessary to feed hay to the cows for eight years. The cows are medium-frame Angus. John said a smaller, easy-fleshing cow was necessary for a no-hay system in northwestern Nebraska.

All replacements are purchased as grown cows and bred to Chianina bulls. These Angus/Chianina cross yearlings are custom-fed and sold to a lean-beef specialty packer in Minnesota. They are considering moving their calving season back to the first of June to allow them to gain more body condition prior to calving. A high level of body condition prior to calving is critical when the cows are being used as a "scavenger" follower herd behind steers. While receiving no hay, the cows are fed cottonseed meal cake as a protein supplement.

Hay is fed to the weaned calves during the winter. Most stocker cattle are purchased in January - March and hay is fed to them until green grass is available. This stocker-quality hay is cut from the meadows in their spring surplus season by contractors for $20 a ton. To avoid negative margins, the Ravenscrofts have been buying mostly terminal sire stocker heifers. These are inter-vaginally spayed by a local vet shortly after arrival for $5 per head. He said his vet can do up to 500 heifers a day.

"Heifers at slaughter usually sell for within a dollar or two of steers, but you can frequently buy them for $10 a hundred less than steers as stockers. If you are willing to own them through the feedlot, heifers can be a better buy than steers," John said.

High Stock Density Grazing Conquers Thistles in Argentina

Argentina has long practiced a system of "pasture rotation" in its humid grassland areas of plowing down permanent pastures every five to seven years and planting them to crops for two to three years. This is done, not for income diversification purposes, but for pasture thistle control.

Under Argentina's traditional system of rotational grazing, using low stock densities and very big paddocks, thistles become a major pasture weed problem. However, Dr. Marcos Giminez Zapiola in Entre Rios Province has permanently parked his disks and plows and now relies exclusively on high stock density grazing in the winter to keep his pastures virtually thistle free. "We thought you had to plow and plant to have good grass. However, I learned in the United States this was not so," he said.

A student of Allan Savory and Bud Williams, Dr. Giminez and his brother-in-law, Manolo Parera, have subdivided the 5000 acre ranch into 500 permanent paddocks with one-wire electric fences. These permanent paddocks are further subdivided into 1300 temporary paddocks with tread-in posts and polywire. All interior gates are made by temporarily lifting the wire with a 10-foot plastic pole. This whole system is powered by just two Gallagher energizers and cost less than $10,000 to put in. Special "training paddocks" with off-set wires are used to familiarize new arrivals with electric fences before entering the main portion of the ranch.

Rotation length (paddock rest) varies from 30 days in the spring to over 100 days in the winter. During the severe 1995 drought a rotation length of 210 days was achieved with a stocking rate of 1956 head thanks to the numerous paddocks. The stocking rate is adjusted to match seasonal forage growth changes by buying and selling, and no stored forage or supplemental feed is used. The ranch normally supports 3300 yearling cattle. These are normally divided into herds of 300 to 500 head.

Crops of linseed, sunflowers, milo, corn and soybeans were once grown as a "pasture rotation" program, but were never as profitable as grazing. All cropping ceased in 1992. "At a stock density of 100 steers per acre, we started seeing the thistles just naturally die out after two years. We don't mow anything anymore." Another byproduct benefit of high stock density grazing (also seen by graziers in Texas) has been a tremendous decline in fire ant mounds. "We never had ant problems until we started cropping. Now that we have stopped and are using high stock density grazing they are moving away."

Dr. Giminez' pastures now contain both warm-season and cool-season species including fescue, Reed's canarygrass, wild oats, perennial ryegrass, dallisgrass, bahiagrass and on his most fertile soils, Matua brome. Legumes include red and white clover, native legumes and Birdsfoot trefoil. Alfalfa, which is traditionally thought of as the base legume in Argentina, does not grow well in the low phosphate, heavy clay soils of Entre Rios. "We call Birdsfoot trefoil 'poor man's alfalfa,'" he said. Phosphate is the ranch's primary soil nutrient deficiency. He said phosphate fertilization was particularly critical for perennial ryegrass survival. "We do all we can with grazing management first and then we start applying phosphate."

With a sale price for the ranch's finished cattle at only 35 to 40 cents a pound, costs have to be kept low. He said the ranch's direct operating costs are only $1.00 per hectare per month (40 cents per acre). "Our production goal is to produce one pound of gain per head per day for 365 days solely from the pasture." He said good Argentine graziers can produce beef gain for between 15 and 17 cents per pound. The Argentine peso and the USA dollar are at par with one another.

His steers are purchased at 400 lbs and typically die at 1000 lbs at between 18 and 24 months of age. Most of these are Brahman/British crosses from the subtropical region of northern Argentina. The ranch's two cowboys are constantly sorting the cattle so the animals nearest to finishing are given the best grass. His Bud Williams training has proven invaluable in reducing stress and shrink in this sorting process. His biggest management headaches are "tail-enders" that refuse to fatten. His steers are sold on the European export market and must grade similar to USDA Select. Marketing is a major cost at 10 to 12% (includes shrink and commission). All slaughter cattle in Argentina are killed on the rail.

Stock water development has been provided primarily by multiple small dams on a small flowing creek that meanders through the ranch. Windmills provide piped water for other paddocks. Stock ponds (rainwater catchments) are used only in emergencies due to their poorer water quality. Having nearby water in each paddock has increased animal performance. "We always noticed that the animals that hung around the water trough all day gained the least." Willows have been planted in swampy paddocks to dry them and to serve as a drought feed reserve. Small trees have been de-limbed of their lower branches to provide a moving, and therefore, healthier shade.

Horses are used as follower grazers to the steers and have proven to provide excellent weed control. The electricity is always turned off in the paddocks the horses are in. A shocked horse will start running and can easily tear down the minimalist one-wire interior fences and could possibly become entangled in the wire.

A small herd of very tame dairy cows is used to graze the access lanes, roadsides and lawn of the hacienda. "We have found the less you plow, the less you spend and the less you spoil the land," Dr. Giminez said.

Gainers and Garbage Eaters Set Record

A leader-follower, 12-paddock management-intensive grazing system allowed the Noble Foundation to produce a record gain of 723 lbs of beef per acre from non-irrigated bermudagrass/annual ryegrass pastures. All annual ryegrass seedbed preparation, planting and some seed spreading were done solely by the concentrated animals. R. L. Dalrymple, pasture specialist for the foundation, describes leader-follower grazing as "gainers and garbage eaters."

Gainer steers were allowed to top-graze the best of the grass sward and then were followed by Brahman-cross heifers that were used to utilize the low-quality grass residue and maintain pasture quality. The gainer steers turned in an average daily gain of 2.2 lbs per day for 168 days and the follower heifers gained 1.1 lbs for 227 days. The gainer steers contributed 544 lbs per acre and the follower heifers contributed 179 lbs per acre. The total pasture cost per pound of beef produced on this system was only ten cents per pound, of which nine cents was for fertilizer.

Pasture Profits

New Zealand consultant, Vaughan Jones, warns there are many times when high stock density grazing should not be used, such as in the following situations:

* Animals are due to give birth and are doing so.
* Avoiding pugging on wet soils.
* Getting early spring grazing when some snow has melted, but not all. Don't overdo the consumption of short lush pasture.
* Grazing new pastures to reduce damage to them.
* Grazing tall crops such as Berseem clover, which has gotten half a meter (20") tall and would be trampled flat under intensive mob stocking.
* Sheep and lambs are grazing some crops. They are better walking through it rather than trampling and wasting it, then going hungry.
* Trying to get rapid weight gains and finish stock. Followers can clean up leftovers, or pasture can be better grazed at subsequent grazings.
* Preventing stockpiled pasture from getting too long. Nipping off the top of the grass with light low density grazing does this.
* Grazing over-long stockpiled pasture. Making stock eat mold in the base of pasture can cause a multitude of animal health problems.
* Grazing old wet pasture with mold in the base.
* Avoiding soil erosion.
* First getting into grazing.

This last point is most important. People who go right from no grazing or extensive grazing one day to high stock density (intensive) grazing the next can be completely disillusioned when weight gains crash, animals are distraught and walk around like crazy, half starved and making mud like you've never seen. Change slowly! Some grazing failures occur when people change too quickly and their grazing becomes too intensive, too soon. Make sure your pastures, animals and you are ready for it.

Chapter 16

Finishing Cattle on Pasture

"The wise man eats to live, while the greedy man lives to eat."
Proverbs 13:25

As a rancher's kid growing up in Mississippi, I had a very limited experience with the grain-finishing of cattle. While we ate pan-fried steak three meals a day, it came like our milk, straight from the pasture.

My Dad, as an experiment in increasing our beef's eating quality, once instructed me to feed a hundred-pound sack of shelled corn to a slaughter-bound spring yearling on ryegrass at the strictly rationed rate of one coffee can of corn per day. Dad said he could taste a big difference in this corn fed beef, but I couldn't. We never repeated this experiment, so maybe he really couldn't either.

Dr. Bill Oliver of LSU told me about a conference at Texas A&M he attended one time that was set up to determine exactly how long cattle had to be on feed to taste good. After several hours of acrimonious debate, an old cowboy sitting with his chair leaned against the back wall, raised his hand, and asked the group if they would like to know the secret to what made beef taste good.

Bill said the room got as quiet as the proverbial mummy's tomb. The old man stood, cleared his throat, and loudly said, "Hunger."

In 1988, the value of American beef production on the farm was $37 billion while the total value of the corn crop was only $9 billion. It has always seemed strange to me that we in the American beef industry have always let the corn-tail wag the much larger dog. To quote Abbot and Costello, "Who's on first here?"

Every time corn gets near $3.00 a bushel, this tail not only wags us but body-slams us vigorously against the concrete. What is really strange about this to me, is not the beating we get, but that so many of us defend this periodic bloodying as being necessary to protect our beef's "quality." This would come as a big surprise to most of the rest of the world.

A recent USDA Farmline analysis of the world beef market noted that most of the grass-fed beef being produced in New Zealand and Argentina from 18-month-old, Angus-cross steers will easily meet the USDA Select grade.

I have eaten both New Zealand and Argentine grass-fed beef and while it is a little more chewy than North American beef it does taste pretty good. In fact, Argentine beef has a much stronger beef flavor than our beef. In Europe, Argentine grass-fed beef is the recognized standard for beef taste and their domestic beef producers try to emulate it.

Youth, the Key to Quality

The primary key to eating quality in grass-fed beef appears to be a slaughter age of 24 months or less. (While there is a market among health-conscious people for very lean beef, it is a relatively small market.) The average consumer will argue over the vagaries of what he wants his beef to taste like, but they all want it tender. Most Americans talk lean and buy fat. Research at North Carolina State found that carcass fat cover prevents a phenomenon known as "cold shortening." This results in muscles that are hard to shear through. As long as grass finished animals have a body fat cover of one quarter to one half an inch of fat, there is no difference in the tenderness between grass finished and grain finished beef.

Easily fattened breeds are Wagyu, Angus, Jersey, Guernsey and Red Angus. Research at Colorado State ranked Wagyu, Jersey,

Guernsey, Angus and Red Angus as also being the breeds that consistantly produce the tenderest meat.

The only breed found outside of the early maturing breeds to produce consistantly tender meat was Piedmontese. Gary Smith at Colorado State said this one breed proved the lie that marbling and tenderness were correlated. Piendmontese will not marble but are very tender. Apparently tenderness is primarily a genetic attribute and depends upon the presence or lack of connective tissue in the muscle. Grain feeding also has little to do with flavor.

Trained taste panelists at NC State were unable to distinguish a difference between beef finished on grazed millet and grain finished beef. Consumer studies have found no preference for steaks and roasts from drylot finished cattle versus grass finished cattle based upon tenderness, taste and aroma.

Young Grass Finished Beef Researched

In the 1970s the general belief was that the world was on the brink of starvation and America would not have the luxury of grain-fed beef much longer. As a result the ARS coordinated a South-wide research project into the feasibility of producing a year-round supply of 18-month-old, 1000 lb, grass-fed steers similar to those produced in Argentina. After three years of research it was concluded that the South could indeed produce 1000 lb steers for at least six to eight months of the year at a less than feedlot cost. The key to this system was to have the animals "finish" on a very high quality forage in the final months before slaughter.

The research showed the South's big production problem in a complete year-round system was the lack of a "finishing" quality forage in the mid-summer and early fall. Such seasonal production "potholes" are a problem everywhere a nearly constant year-round supply of beef is needed.

While not thought of as grain-fed beef countries, Argentina and South Africa, do feed grain in the finishing stage during the gaps in their forage flow. Grain is not fed in the growing phase during these poor production periods since the use of compensatory gain on a subsequent higher quality forage is much cheaper.

In North America, North/South regional production partnerships could help fill the holes in a year-round system. However, what really takes the steam out of grass-fed slaughter beef in the USA is that our slaughter price is not normally the highest per pound price.

For example, at the peak of the last cattle cycle an 800 lb grass-fed steer sold for $83.50 cwt as a feeder animal, but only $75 as a Choice grade grain-fed slaughter animal. Even 1000 lb grass steers were bringing $78.50 as feeders on the $75 fat market.

As long as domestic cattle feeders are willing to pay such premium prices for feeder cattle and work on paper-thin margins, they will have little to fear from domestic grass-fed slaughter beef production. However, if we see a situation where the slaughter price becomes the highest class price as it was in the summer of 1996, the whole scenario quickly changes. This will require a grain price consistently above $3.00 a bushel.

In the short term, it appears the best approach for most of us is to keep working on extended stocker programs that will allow us to take our cattle to heavier and heavier weights before they are placed on feed. This approach requires no new markets or customers and would help us meet our biggest threat — domestic grain-fed chicken — head on.

Consider the fact that a grass-grown 800 lb steer that is grain-fed to 1100 lbs has approximately the same grain-to-slaughter weight conversion ratio as a chicken. At a feeder weight of 850 lbs, we exceed chicken's efficiency. At 900 lbs we exceed the slaughter weight-to-grain conversion ratio of catfish! The next time you hear Frank Perdue running off at the mouth about his chicken's feed efficiency, quote some of the above figures.

Several years ago scientists at the University of Tennessee found that a one pound of grain per pound of gain conversion ratio (1100 lbs of grain for an 1100 lb slaughter steer off forage at 950 lbs) produced a quality beef product acceptable to 100 percent of those who sampled the beef in the university's taste tests.

In beef's battle with the big bird, grass is our ultimate weapon.

Grain-on-Grass Cattle Finishing

There is starting to be a revival of interest in grain-on-grass cattle finishing systems. These systems were widely researched in the South in the mid-1970s, and found both cost-effective and capital efficient in the production of Select grade slaughter beef during the cooler months of the year.

During 154 days of wintergrazing on stockpiled fescue at Virginia Tech 1976-77, yearling steers fed 1 percent of their body weight in grain (i.e. nine pounds of grain in a 900 lb steer) gained 2.93 lbs and when swapped to bluegrass/white clover in the spring gained 2.41 lbs per day and finished in the high Select range.

Another group of yearlings were fed at 0.5 percent of body weight until the last 90 days and then fed at the one percent level until slaughter. These steers finished in the medium Select grade.

Yearling steers finished solely on grass averaged 1.44 lbs and finished in the high Standard grade. It was this tendency to drop into the heavily discounted Standard grade that made grass-finished cattle such a drag on the late 1970s cattle market.

Attempts to salt-limit grain consumption so that daily feeding would not be necessary produced highly variable average daily gains and were not recommended. Also, grain supplementation during the spring lush period was found not to be cost-effective as spring bluegrass/white clover pasture could produce daily gains equivalent to a high grain ration.

A one percent level of grain supplementation was found to replace as much as 50% of the grass intake of the cattle and allowed more steers to be overwintered per acre of stockpiled fescue. This made better use of the spring pasture surplus. Virginia Tech recommended that grain supplementation not be started on stockpiled fescue until January as the late fall and early winter quality of stockpiled fescue alone was adequate for good animal gains.

Summer Finishing Expensive

Grain fed at the one percent level allowed very heavy summer-finished steers to continue to gain at an acceptable 1.59 lbs per day although at a far higher grain cost compared to cool-season gains. The grain to gain ratio over grass alone increased to nearly 20 to 1 in the summer over non-supplemented bluegrass/white clover. At a nickel a pound grain cost that would equal one dollar of grain for each pound of additional gain produced.

Similar high costs were produced on grain supplemented, warm-season grass, summer pasture in Mississippi and Louisiana as well. Perhaps, the best advice would be to avoid finishing cattle on pasture during the hot summer months.

Keep in mind that low roughage/high grain "hot rations" produce less heat stress on cattle than high roughage rations. The whole-shelled-corn programs of Moorman's, Tend-R-Lean and others produce good summer gains even in the steamiest parts of the South.

Grain on grass programs should be analyzed "holistically" including their ability to balance pasture seasonality, lack of need for protein input, low capital requirements, natural manure recycling, and the increased marketing and production flexibility for their true potential worth to be seen.

Grass Cheaper Than Chicken Manure

In the Deep South, there has been an effort to bail the broiler industry out of its growing environmental problem by promoting broiler litter as a "cheap" cattle feed. However, recent research at Auburn University's Sand Mountain Substation pointed out that graziers need to be skeptical of such propaganda.

Research by D.I. Bransby and J.T. Eason showed a **decrease** in profit per acre in excess of $100 per acre when stocker cattle on both infected and non-infected fescue were supplemented with a "cheap" ration of 50:50 broiler litter and corn.

Non-supplemented steers stocked at three per acre on infected fescue on a fall to spring graze returned a profit of $290 an acre. Non-supplemented steers stocked at two per acre on non-infected fescue showed a profit of $408 an acre!

Supplemented steers stocked at the higher rate of 3.5 on infected fescue returned $180 an acre, a decrease of $110 an acre. Supplemented steers stocked at 3.5 per acre on fungus-free fescue showed a decrease in profitability of $143 an acre! To paraphrase the Argentinean graizer's comment about cheap corn, "The chicken poo-poo may be cheap, but the grass, she is always cheaper."

Grass Finished Beef Argentine Style

If you want to learn how to finish cattle on grass, go to Argentina. Argentine grazier, Marcos Giminez-Zapiola, told me that the whole culture of the country exists to "Put a great steak on your plate."

One of the first things I learned about Argentine grass finished beef is that it is not extremely lean. The average yield is 60 percent or not appreciably lower than our grain-finished carcasses. The USDA estimates that most Argentine beef would grade USDA Select or better.

The Argentines prefer to graze early-maturing Angus-based cattle so that they will fatten and finish before the second winter. Like Gordon Hazard, they prefer to summer the animal twice but only winter it once in its life. Animals the closest to slaughter are given the best pasture. Steers in the growing and framing phase are used as "followers" to the "finishers."

The finishing zone lies at the same latitude as Virginia and Kentucky, receives 38 to 45 inches of rain, has only 180 frost-free days a year, but deep snow is rare. Permanent pastures are alfalfa mixed with fungus-free fescue, orchardgrass, wheatgrass or Matua brome. Red clover and annual ryegrass are overseeded to fill holes in the pasture sward.

Argentine rancher, Pedro Landa, said he likes a 70% grass, 30% alfalfa mix for maximum animal performance and minimal bloat problems. The Argentine tradition has been to rotate between permanent grass and annuals on a seven-years-in-alfalfa/grass pasture and three-years-in-annuals basis. The pasture rotation helps greatly with thistle control which is Argentina's primary weed problem. Often a summer crop of soybeans is grown as a summer annual in one of these years and sold as a cash crop. A typical summer annual rotation would be corn-soybeans-sudan. These fields are then planted to winter annuals for cool-season grazing.

Often one of these is wheat with the cattle moved off early to allow a cash grain harvest. Triticale is the preferred grazing annual because it can be planted early in the dry of late summer and provides the earliest fall grazing.

Corn is stagger-planted and used to graze animals scheduled to "finish" in the summer. Open pollinated varieties are preferred for grazing due to their greater drought tolerance and slower maturity. The sudangrass is used both as a shade smother crop to remove warm season grasses such as common bermuda and as a grazing crop. Grazed oats are used to finish animals in the autumn.

At Hacienda La Chita, a 12,000 acre ranch I visited, all crop and forage planting is done in a modified no-til method whereby all three operations — disking, planting and packing — are done in one pass. Thanks to the smothering effects of the annual crops, no herbicides are needed. At La Chita, a large wall chart shows the major paddock subdivisions. A plastic overlay on the chart allowed the manager to keep up with what crop went where and when.

Hay and/or silage is made from the spring pasture surplus and is primarily fed in the late summer and early fall to allow the pastures time to "stockpile" growth for winter grazing. The most difficult time to produce finished cattle on the Argentine system is in the late winter and early spring. It is only during this one period of the year that the Argentines believe a feedlot does a better job than grazing in producing quality beef.

English Cattle Predominate

Their ideal slaughter weight for an Angus cross is 900 lbs but they will take them as heavy as 1000 lbs and will drop down as low as 700 lbs if the kill supply is short. Animals are sorted for slaughter based upon the fat cover on their rumps. In order to finish before their second winter, English-breed cattle are never allowed to gain less than three-quarters of a pound per day and must gain between 400 and 450 lbs in a year.

In the subtropical north of Argentina, Brahman (Zebu) cross cattle necessarily predominate. These cattle will not finish before the second winter due to their slow genetic maturity and so are often used as follower cattle to the premium priced English cattle. They are roughed through the winter at a very low rate of gain and are allowed to make maximum use of compensatory gain the following spring and summer. Argentine packers do not consider the Brahman cross to be "finished" until it reaches at least 1250 lbs.

A major problem for Argentine graziers are animals that continue to grow and will not "finish" before the third winter of their life. These animals have to be taken into a third winter at a very heavy weight and are very grass inefficient.

The third class of "finishing" animal are Holstein steers. These are used as third grazers and are considered a "scavenger" class of animal. They are traditionally killed in the 1400 to 1500 lbs weight range for export to Europe. Most ranchers graze these as "load-up" cattle on a contract basis.

There is also a small premium-price market for pasture veal from overly-fleshy weaned calves. These calves range from 300 to 500 lbs in weight. Since most Argentine ranchers retain ownership to slaughter, high weaning weights are not prized and calves are traditionally weaned early in life at four to five months of age.

Gauchos ride the pastures daily and sort off animals ready for slaughter. Thirty, 1000 lb steers are a "load" for the small Argentine trucks. This constant "topping" prevents cattle bunching and helps stabilize the prices. The highest prices for beef are in the winter and some ranchers will grain supplement cattle on winter annuals so that they will fatten for this premium priced market. There are also a few North American style feedlots (80% grass silage/20% grain) but they have been found to only be profitable during the premium-priced winter season.

One common element I have found in studying grass finished beef systems around the world is a wide range in slaughter weights. This is apparently necessary to come up with consistent daily slaughter supply.

Build Your Grazing System Backwards from the Goal

Argentine rancher, Pedro Landa, said a grass finishing program must be built backward from its endpoint rather than forward from wherever you happen to start from. "If you change the endpoint to a grass finished animal, it will change your choice of genetics, your calving season, everything. A production system must be a complete whole where one step naturally leads to the next for it to be successful."

Pasture Profits
* Youth is the key to grassfed beef quality.
* Quality grassfed beef is seasonal.
* Grain on grass can increase quality grade and stretch grass supply.
* Avoid finishing cattle on pasture in hot weather.

Chapter 17

Low Cost Grass Silage Production

"A wise youth makes hay while the sun shines, but what a shame to see a lad who sleeps away his hour of opportunity." Proverbs 10:5

It is more profitable to turn your grass into beef, and buy-in your hay needs than to restrict grazing. In a seasonal grazing operation forage conservation programs are primarily used to: (1.) maintain pasture quality by keeping the grass young and vegetative through the removal of overly mature or surplus material and (2.) to balance out the yearly forage growth fluctuations and provide insurance against drought and other unexpected weather events. Low-cost grass silage is an excellent way to accomplish both of these goals with cool-season grasses.

Grass silage has the benefit of needing minimal machinery and labor input compared to hay (still best to hire it done). It also lends itself well to self-feeding. The primary drawback to grass silage is that it is not economically portable over long distances, and therefore not saleable as a cash crop.

Hay is a better rumen developer and feed in very light calves (less than 350 lbs) than silage, but the quality of the hay needed with these very light calves is such that it should probably be purchased rather than made from pasture surpluses.

There are many ways to make grass silage. My book **Quality Pasture** presents more details on this subject. However, since the primary thrust of this book is direct grazing, I will only describe the lowest cost methods of making and using grass silage.

Rumen Change Necessary to Digest Silage

Most of the disappointments with grass silage as a supplement to pasture have been the result of a too sudden change to silage. This shift must be done gradually over at least a week's time to allow the cows' rumen micro-flora to change. Don't wait until you are completely out of grass before you start supplementing with silage.

Keep in mind this gradual transition is also necessary when shifting cattle back to direct grazing after having been on a full silage feed.

Ensiling Quality and Digestibility

A major myth is that ensiling a forage crop improves its digestibility and quality. The truth is that no stored forage is any better than the day you cut it, and always loses part of its quality in the translation.

The primary decider of your silage quality will be the stage of maturity at which it is cut. Length of chop, silage additives and weather, all pale to insignificance compared to cutting the grass before it makes a seed head. Too mature silage can actually consume more energy than it produces, and actually cause your cattle to lose weight.

While often described as relatively weather-proof, silage should not be cut in wet conditions, and it shouldn't be cut before 11 a.m. (Daylight Savings Time) because the sugar content decreases overnight and increases again with sunshine. Low sugar, low energy silage is poor silage, and silage made from heavily nitrated grasses will make high nitrate, low energy silage.

Rain falling on silage does nearly as much damage as rain falling on hay, and must be dried off before ensiling. Never ensile wet material.

Machinery for Silage Making

Grass silage is an easily consolidated crop. The precision chopping required for corn and drier forages is not needed with grass silage as long as clamps or bunker silos are used. In fact, if the grass is chopped too short the animal cannot regurgitate it and chew its cud, and therefore not get the full benefit of the feed. With grass silage, it is now recommended that the cut be as long as possible to increase cud chewing and have the grass remain in the rumen as long as possible.

Low-cost, durable, direct-cut flail harvesters are used all over New Zealand for the making of grass silage. Research at New Zealand's Ruakura Research Center has shown that flail direct cut silage can actually give better weight gains than chopped, prewilted silage, as long as the grass was relatively mature and chopped after 11 a.m. Economic research in the United Kingdom found that these low-cost flail harvesters were the only cost-effective method of forage preservation for small livestock producers and allowed a one-man, one-tractor operation.

Two major drawbacks to a flail harvester are a tendency to suck up dirt and manure along with the chopped grass and accidentally chop into the soil surface on uneven ground. Getting soil mixed in with your silage will cause poor fermentation and should be avoided if possible. To avoid this problem many graziers prefer a chopper that uses a pickup reel to feed the grass into the chopper rather than the vacuum suck system the flail chopper uses.

With wet, sappy grass/legume spring pasture, a higher quality silage can be made by first cutting, swathing, and field drying the grass to the point where moisture no longer runs out when the grass is squeezed in your hand before chopping and packing. This pre-wilting cuts hauling costs by reducing the weight of the grass, reduces effluent losses and potential water pollution, and results in a higher dry matter silage that in some studies has produced a slightly higher average daily gain.

Care must be taken to prevent the cut grass from becoming too dry. If too dry, the grass will become difficult to consolidate and result in moldy silage, which can be toxic to the cattle.

In Holland and Germany, self-loading forage wagons that combine the chopper unit with the wagon are used quite successfully, and reduce the capital investment needed to make small amounts of silage. While touted as one-man silage making units, most grass farmers have found having an extra tractor and man to help with the mowing and later with the spreading and consolidating of the grass at the clamp greatly speeds up silage making.

Dutch graziers have also found that silage making is easier and quality is higher if several smaller silage clamps that can be filled and consumed rapidly are made rather than one big one as is common in Britain and North America. Hauling chopped forage over one-half mile to a silo greatly reduces the efficiency of a low capital, one-forage-wagon operation. By keeping hauling distances to only 1500 feet, the Dutch figure 10 to 18 tons of silage can be put up an hour with a single self-loading forage wagon. Most Dutch silage clamps are no more than 50 tons in size or approximately one day's work.

In fact, if hauling distances are kept super-short, the forage wagon can be done away with, and all the hauling done with tractor-mounted buck rates, a method well illustrated in a book by Newman Turner called **Fertility Pastures**. The tractor-mounted buckrake is probably the ultimate in low capital forage preservation.

Building a Silage Clamp

The ultimate low-cost silo is the silage clamp. All it requires is a space of flat ground with a slight slope to prevent water logging. Concrete pads allow mud-free, self-feeding, and prevent soil contamination from possibly muddy tires of the spreading and consolidating tractor.

Clamps are built using what the British call the "Dorset Wedge" method. The first day's cut is built into a tepee shape with a slope of no more than 20 degrees on all sides. Some graziers leave the sides of the wedge steep as is shown in the Turner book. It is not necessary to consolidate these sides to prevent deterioration as long as the clamp is covered with plastic, but these steep sides are dangerous due to the possibility of tractor turn-over while spreading and consolidating. For safety's sake, it is recommended that the sides also be sloped at no more than 20 degrees.

Your first day's clamp will look sort of like a squashed haystack. If the clamp is to be self-fed it should not be over six feet in height.

The Dutch will seal this one-day clamp with plastic held tight with a layer of dirt or barnyard manure, and build a new clamp from scratch the next day. The British and New Zealanders will frequently add the next day's cutting to the same clamp, although the British admit the Dutch method produces a superior silage.

Grass silage is naturally heavy and needs very little tractor consolidation to make it air tight. In fact, too much consolidating activity can create a "bellows" effect, which actually pumps air into the clamp. Let the weight of the grass do most of the consolidating and save yourself tractor time and fuel.

Vaughan Jones recommends that the clamp be covered with plastic and close-spaced tires each night. He said the rising warm air caused by the ensiling process causes cold air to be drawn in and can cause deterioration and mold.

The next day the tires are removed, the plastic is rolled back, and that day's cut is then piled and packed against one side with a minimum added depth per day of three feet. The 20 degree slope is again replicated. Each day that day's cut is packed against the filling side. When the clamp has been completed, it will look like a soggy loaf of bread that has been dropped and has squished out on the sides.

The real secret to successful grass silage is quickly sheeting and weighting of the cover to prevent wastage. If this is not done quickly or done well enough, your whole exercise will have been a waste of time. Vaughan Jones said to remember that cut grass continues to deteriorate as long as it is allowed to breathe. The clamp must be air tight to stop deterioration.

The covering plastic can be covered with closely-spaced tires or even better six inches of dirt or barnyard manure. The dirt will prevent the silage from freezing in very cold climates. If the plastic is held tightly against the silage wall, there should be no deterioration and wastage at all. Research in New Zealand found the silage from a well-made and sheeted clamp to be statistically equal to that of an American upright, air-sealed metal silo. Seeding the covering soil and/or manure to a quick growing grass will prevent erosion.

The clamp should be fenced off to prevent cattle and deer from climbing on it, and a shallow trench should be plowed around the site to prevent waterlogging. Field wilting prior to chopping will reduce effluent runoff to virtually nothing, however care should be taken to capture silage effluent both for reuse and to prevent stream pollution. Research in Northern Ireland has shown that stocker cattle will readily consume up to three gallons of effluent per day, and that the effluent has about the same feeding value as barley. Some Irish graziers line the bottom of their bunker silos with two layers of straw square bales to soak up the effluent for later feeding. Others capture the effluent and spray it on their paddocks as fertilizer.

Feeding Out and Supplementing Grass Silage

Stocker cattle that are allowed to feed to appetite on grass silage will seldom gain over 1.5 lbs per day and more often gain at half this rate per day. This rate of gain is acceptable as long as the cattle are being wintered for subsequent spring grazing. However, if the cattle will be sold before spring grazing, a higher rate of gain can be produced by supplementing with a small amount of grain and protein.

One British research study found that supplementing stocker cattle on a full feed of grass silage with one kilogram (2.2 lbs) of barley produced an average daily gain of eight-tenths of a pound of gain per day. This gain was increased to 1.98 per day with a 40% barley/60% soybean mix and increased to 2.09 per day when a 75% barley/25% fish meal mix was used. The latter two mixes were still only a kilogram per day total feed.

Such responses to protein are only found in younger cattle. The higher gain from fishmeal was thought to have come because fish meal is only slightly digested in the rumen and becomes a source of bypass protein.

The British have also found that the higher the quality of the silage the less impact grain supplementation had on increasing average daily gain. If your goal is to maximize the use of your home-produced silage and minimize purchased grain, it is seldom wise to feed over two to three pounds of grain/protein mix supplement per day. This level of feeding should be adequate to produce an average daily gain of two pounds a day.

Self-Feeding Silage

A one-wire electric fence must be used to keep the cattle from climbing on the silage stack if the silage is to be self-fed. To prevent waste and allow feed rationing, push solid fiberglass posts with an electric wire attached horizontally into the silage face three feet above the ground. This electric feed bunk will allow the cattle to eat forward through the silage stack until they touch the wire. To feed the cattle, all one has to do is push the posts deeper into the silage.

A minimum of eight inches of silage face per animal is necessary to insure a full daily feed. There is some silage waste using this method of feeding but the waste and manure mix makes good bedding for the cattle and will produce heat as it self-composts both important factors in areas with cold winters.

Permanent concrete pads can keep the silage face mud-free but require that the manure be captured and respread to prevent severe nutrient transfer from occurring. British dairyman, Newman Turner, moved his self-fed silage stacks across his pastures from year to year and used the intensive manuring and pugging as a way of improving the tilth and fertility of his soils.

Self-feeding lends itself well to herds that will be totally off of pasture for an extended period of time. There are no tractors to crank and no need for feed wagons. The primary problem is that your manure and nutrients are concentrated in one very small area and severe soil nutrient transfer can occur.

Pasture silage has long been the winter feed of choice for

Alberta beef cattle ranchers Cecil and Carol Hoven. However, for the last two winters, the Hovens have been letting the cows do most of the work involved in winter feeding by using self-feeding free-standing silage clamps. "Self-feeding saves us the two hours a day we used to spend hauling the feed to the cattle. Now we bring the cattle to the feed and let them feed themselves," Cecil said.

The Hoven's ranch is located approximately halfway between Calgary and Edmonton and is north of the largely snow-free "Chinook" zone. Snow can cover the ground at the Hoven ranch for up to six months of the year. However, thanks to an aggressive pasture stockpiling program they have been able to cut their winterfeeding time to no more than 125 days. "This year our goal is to cut our winterfeeding to 100 days," Cecil said.

The Hovens have 350 cows and overwinter all of their calves. To do this requires two clamps of pasture silage. Separate clamps are provided for the cows and the weanlings. These are wall-less freestanding clamps 100 feet wide by 150 feet long. The 100 foot width was chosen to match the width of the plastic sheeting available. The clamps are built on dirt. The silage is made from orchardgrass, fescue, red clover and alfalfa mixed sward pastures.

The clamps are approximately six feet high and are flat on top. Straw is used to keep the plastic pressed tightly against the silage and to prevent the silage from freezing. The clamps are built in an opening in a copse of trees. The trees provide a natural windbreak. The cattle have no access to any roofed or closed structure at any time. (John and Bunny Mortimer detail the use of trees with your livestock in their book **Shelter & Shade, Creating a healthy, and profitable environment for your livestock with trees**.)

The cattle feed from both sides of the silage clamp simultaneously and the feeding is regulated with a one-wire electric fence strung from poles driven horizontally into the silage face. To feed the cattle one simply taps on the end of the fence post with a shovel or pickaxe. This drives the post deeper into the silage. The cattle can then eat into the silage until they are stopped by the electric wire. The Hovens do this once a day. The calves receive 2 to 3 pounds of grain every day in addition to the silage. This is fed on the ground underneath a one-wire electric fence.

Feeding from both sides of the clamp simultaneously is important in very cold climates as the side of the face toward the wind can freeze when the temperature drops below -20 F. "This is not as big a problem as many people think it would be. Only the side of the face toward the wind will freeze. We have found this to happen only two to three times a winter."

A frozen silage face requires a few minutes work with a tractor and front-loader. The silage seldom freezes over an inch or two into the open face. Carol Hoven said the secret to self-feeding on dirt with no mud is to have the silage consumed before the spring thaw. "We have found the best use of stockpiled pasture is in the spring more than the winter. We are grazing by the first of April and are already on pasture when the thaw occurs," she said.

Rolling "tumblewheel" electric fences are used to stripgraze the stockpiled pasture. These eliminate the need to drive posts into frozen ground. Cecil said they have found no problems in getting the cattle to graze through the snow as long as there was good feed beneath it. "Our stockpiled pasture still tests at twelve and a half percent protein the following spring. In other words, we've got the same quality as good hay, but at half the cost," Cecil said.

Salt Added to Silage Clamp Helps Prevent Freezing

Some northern graziers have been using most costly and higher weather risk, wilted silage to avoid silage clamp freezeups. Gaylord, Minnesota grazier, Doug Gunnink, gives the following tips on how to prevent direct-cut silage clamps from freezing.

☆ The addition of 3 lbs of Solmin mineral mix and 4 lbs of salt per ton of wet silage plus an additional 100 lbs sprinkled on the top of the clamp prior to sealing has helped keep silage clamps from freezing. Apparently, adding salt also makes the silage taste sweeter to the cattle and can increase consumption. (This is the same taste effect you experience when you sprinkle salt on a watermelon and for the same reason, to make it taste sweeter.)

☆ Locate the clamp so the southern end can be opened. By feeding with a southern exposure some solar energy will help prevent the silage face from freezing.

☆ Use black plastic and leave about one foot to hang over the edge so the pile catches more sun, but also retains the heat of the ground and silage.

☆ High-sugar-content forages do not freeze as readily as those with low sugar content. Keeping soil calcium levels high and only cutting silage after noon helps produce high-sugar-content forages.

Swath Grazing

An alternative to both hay and silage that I have seen used in low humidity Western Canada is to cut and swath your upright hay-type grasses in the late fall and leave them in a windrow. The forage inside these windrows stays green and of stocker cattle quality all winter long, allowing direct grazing in relatively deep snow. Temporary electric fences are necessary to prevent the cattle from sleeping on and wasting the windrow. The beauty of swath grazing is that the cattle are brought to the feed, and manure is naturally spread over the paddock.

Canadian graziers have found these swaths to be more reliable than pasture stockpiling. While cattle readily eat through snow to get to stockpiled grass underneath, they won't do this through a layer of ice. The swaths have been found to always be easy for the cows to break into regardless of the ice conditions.

I do not know how well this would work in cold areas with high humidity. I know it would not work well due to mold buildup in areas that have warm and wet winters.

Pasture Profits

* Grass silage is a cheaper, more dependable method of capturing pasture surpluses than hay. With small amounts of additional grain, quite acceptable gains may be produced.

* As with all stored forages, care must be taken to prevent severe nutrient transfer either by taking the manure to the pasture or the silage to the pasture.

Chapter 18

The Ultimate Skill —
Management-Intensive Grazing

"Lazy people want much but get little, while the diligent are prospering." Proverbs 13:4

Management-intensive grazing can increase your per acre production between 20 and 40 percent. This increase will come either as increased animal production and/or increased hay production. After the initial subdivision expense it is almost a pure reward to management. This is why management-intensive grazing is becoming the skill that separates the professional grazier from all the rest.

Benefits of Managment-Intensive Grazing

While per acre increase in stocking rate and gain are usually listed as the primary benefit of management-intensive grazing, Keith Milligan, senior pasture extension officer in New Zealand, gave me a longer list.

1. A better return on total investment through a higher stocking rate, increased per head production, and lower death losses from better animal observation due to animal bunching.

2. A lower labor input due to a more even year-long work

load without high peak periods due to massive haying or feeding.

3. A general conservation of the environment due to less over-grazing, better utilization of rainfall and fertilizer due to faster pasture cycling, and the ability to preserve important preferred species of grass.

4. A much increased sense of peace of mind on the part of the grazier. He can see his feed bank out ahead of him, and by measuring the grass' regrowth can make buying and selling decisions far in advance of an actual "crunch."

It will be very hard for traditional producers to compete with a grazier who has a 20 to 40 percent production advantage.

First Subdivision

In the humid regions, the first subdivision should provide for a sacrifice area. This is an area where the animals can be placed and fed hay during periods of high rainfall. A sacrifice paddock is particularly critical with annuals and alfalfa to avoid plant damage. A sacrifice paddock can be a pasture with a tight sod or a trap with three or four feet of sawdust or wood chips laid over it.

Fresh Water

If you are currently watering your stock out of a pond or dirt tank the cattle can get into, put your money into a fresh water system before tackling management-intensive grazing. Management-intensive grazing will only make this very bad animal health situation worse. If you won't drink it, you shouldn't force your stock to drink it. Dirty stock water is your primary source of disease and parasitism.

Permanent Subdivision

250 paddock subdivisions are not 50 times better than five subdivisions in environments that produce rapid grass growth and, in fact, could be fifty times worse. In humid environments, including irrigated pastures, a very large number of permanent paddock subdivisions may allow the fast-growing grass to become mature and stemmy before the grazier can get to them.

Temporary Fences

In high rainfall environments, 90 percent of the benefit of management-intensive grazing can usually be accomplished with as few as five to ten permanent subdivisions, plus the use of temporary fence during slow growth periods.

Note, I said plus the use of temporary fence.

During winter, a ten-paddock cell may be easily subdivided into 100 or more paddocks through the use of movable, one wire electric fence.

Begin in Slow-Growth Months

New Zealand dairy farmer, Ray Metcalf, said the sole intent of a grass plant is to make a seedhead, and the primary task of the grazier is to prevent it from doing so.

Consequently, I do not recommend for anyone just getting started with management-intensive grazing to begin in the spring season with cool-season grasses or during the mid-summer season with tropical grasses. Concentrate your efforts on the slow growing months, or even better, on winter months when the results of your efforts can be dramatic, and the chances of your accidentally hurting animal performance will be low.

The practice of stockpiling fall-grown cool-season grass and then tightly rationing it out over the winter as a replacement for, or a protein supplement to hay has the highest profit return of any management-intensive grazing program I have seen. This would be an excellent starting point.

Flexible Stock Policy

Milligan said that while intensive grazing could raise per acre production by 20 to 40 percent, management-intensive grazing in conjunction with a flexible stock policy could increase production by 60 to 80 percent!

A flexible stock policy changes the stocking rate to match the seasonal variations in grass growth. This flexibility can be accomplished by buying and selling animals, but this is extremely risky because the gain per head is likely to be too small to absorb the price rollback.

It would be much better to sell gain or to have a forage sequence planned that matches the growth of the various forages. For example, on spring fescue you could have a set of light calves that would be summer grazed and finished on fall pastures, plus an overwintered set that would be shifted to cowpeas or millet for a summer finishing program.

This would "load-up" the fast-growing spring pastures and increase both the per-acre and per-head performance of the cattle. The chapter on using compensatory gain has some examples of this forage sequence planning.

Management-intensive grazing helps immeasurably in this flexible stock policy in that there are few pasture surprises. By closely monitoring the grass regrowth, you can see a drought shortage coming weeks before your neighbors. The same is true for the start of spring. This can help you buy and sell cattle before the rest of the crowd figures out what is happening. Monitoring the grass also tells you how the cattle are gaining.

Haying and Silage Programs

The alternative to a variable stocking rate is a haying or silage program that takes grass from surplus periods such as the late spring to deficit periods such as the winter. This should not be thought of as a replacement for a planned forage sequence of cool-season and warm-season grasses but as an insurance policy.

When you are just starting out with management-intensive grazing, plan to take the increase in forage off as hay or silage in your early learning years. Once a comfortable hay insurance reserve has been built, you can start edging up your overall stocking rate. Keep in mind that with stocker cattle and management-intensive grazing, haying is primarily done to maintain pasture quality rather than to produce hay.

Manipulating Grass with Livestock

One major disappointment of many new intensive graziers is a lowered per head production. You must keep in mind that high production per animal and high production per acre are opposite ends of a teeter-totter. If one goes up the other must go down.

For example, the potential for a high average daily gain is determined by how tall the grass is when you turn the cattle on a paddock. The actual average daily gain is set by how much grass is left when the cattle leave the paddock. Grazing the grass right into the ground so as to leave a sharp graze line between paddocks before shifting produces a low average daily gain.

By leaving a lot of grass and only allowing the cattle to top graze the best of the grass sward, average daily gains equal to those of a full feed of grain can be achieved.

I point this out not to suggest it is cost-effective for animals to be managed for high rates of gain, as it is very wasteful of the grass, but only to illustrate that it can be done. Young, tender, green grass leaves are the ultimate concentrate for ruminant animals.

The only way to break the per head vs. per acre teeter-totter is to use high production animals as first grazers and lower production animals as last grazers. By understanding and knowing how to manipulate grass with the cattle, a grazier can control both his animal's gain performance and the per acre production to a remarkable extent.

Your Grass Eye Will Come

Training your "grass eye" just takes time and practice and it will come. However, I do not recommend you start learning management-intensive grazing with high production animals such as lactating dairy cows, stocker cattle or weaned lambs. Learn with replacement heifers, beef cows and calves, or dry dairy stock that have lower production goals.

Also, it has been my experience that perennial grass pastures need about three years of "conditioning" by way of management-intensive grazing before they are ready to really turn a flip for you with the high production animals and classes anyway. So give yourself some time to learn.

In most cases it will take 10 to 12 years of grazing experience for you to truly become adept at management-intensive grazing. If you aren't willing to make this commitment, you may want to consider another line of work.

Pasture Profits
If You're Just Starting Management-Intensive Grazing
* Start in the slow or no growth months.
* Do not increase current stocking rates. Prepare to make hay instead.
* Use low production animals.
* Remember, Rome wasn't built in a day.

Epilogue:

And Now, the Really Hard Part

"It is senseless to pay tuition to educate a man who has no heart for the truth." Proverbs 17:16

A lot of you may still be unconvinced that directly harvesting a grass crop with animals is the way to go. This probably means you don't have enough information, or you may need to try some of these ideas on a small scale first to get that belief, or find a mentor to coach you along. Do whatever it takes to convince yourself you are doing the right thing before you put a lot of money into stocker grazing.

You will not be successful with grazing or any other venture until you first believe you can be successful. A little skepticism is healthy but too much will virtually guarantee failure. Remember, no one ever walked a rope across the Grand Canyon who didn't first believe he could walk a rope across the Grand Canyon. It is finding that belief in one's self that is the most difficult part of grass farming and life itself.

A great many people come to grass farming after having crashed and burned in some other agricultural endeavor. As a result of a previous failure, these people have very little faith in farming or themselves. It is easy after a major failure in life to fall into the trap of seeking out co-misery.

You've seen these people. They're in every rural coffee shop in North America. Up in Canada they call them pity parties. Farmers who feel rotten about themselves routinely join together to moan and groan and gripe about the hopelessness of it all. As bad as they feel, the one thing that makes them feel worse is for a farmer to become successful through his own initiative. Their primary role in life is to wage psychological warfare on anyone showing any threatening signs of positive thinking. They know that whenever we attempt something new we will always do things wrong while we are learning and will look and feel pretty foolish. And that's where they will concentrate their attack.

The "I told you so" chorus will be long and loud and the ridicule unceasing. And this usually works. Just one negative person can sap the positive energy out of a large group of positive people.

The only defense you have against energy-sapping negative thinking is to get as far away from negative people as possible. Seek out the positive people in life. They're out there. Haven't you noticed that the people you meet who are successful in life generally have an optimistic outlook on everything? I guarantee you, this positive outlook preceded their success.

Remember, people who don't think they can succeed, don't. For anything to work out for you in the pasture, you must work it out first in your head. Here's a little parting word of advice that may make life easier for you.

Take Care with the Company You Keep

☆ You can't teach a pig to fly. Not only is it a waste of your time but it annoys the pig.

☆ Don't make it tougher on yourself, possibly too tough, by trying to drag a bunch of squealing, resisting pigs up the success ladder with you.

☆ Peer pressure is designed to contain anyone who shows a sense of drive. The best way for unsuccessful people to cope is to try and make sure no one is successful. You can't afford to be popular with negative people.

☆ Surround yourself with people most like the person you want to become. You always become more like the people you associate with and less like those you don't associate with.

☆ Success is indeed contagious, but so is failure. Take care with the company you keep.

☆ Good grazing and a good life to you.

Glossary

ADG: Average daily gain.

AI: Artificial insemination.

Aftermath: Forage that is left or grown after a machine harvest such as corn stalks or volunteer wheat or oats. Also called the "Fat of the Land."

Animal unit day: Amount of forage necessary to graze one animal unit (one dry 1100 lb beef cow) for one day.

Annual leys: Temporary pastures of annual forage crops such as annual ryegrass, oats or sorghum-sudangrass.

AU Animal unit: One mature, non-lactating cow weighing 500 KGs (1100 lbs) or its weight and class equivalent in other species. (Example: 10 dry ewes equal one animal unit.)

AUM: Animal unit month: Amount of forage needed to graze one animal unit for a month.

Blaze graze: A very fast rotation used in the spring to prevent the grass from forming a seedhead. Usually used with dairy cattle.

Break grazing: The apportioning of a small piece of a larger paddock with temporary fence for rationing or utilization purposes.

Breaks: An apportionment of a paddock with temporary electric fence. The moving of the forward wire would create a "fresh break" of grass for the animals.

Cell: A grouping of paddock subdivisions used with a particular set or class of animals. During droughts, several cells and their animals may be merged and operated as one very large cell and herd for rationing purposes.

Clamp: A temporary polyethylene-covered silage stack without permanent sides or structure made in the pasture.

Continuous grazing or stocking: Allowing the animals access to an entire pasture for a long period without paddock rotation.

Coppice: Young regrowth on a cut tree or bush.

Compensatory gain: The rapid weight gain experienced by animals when allowed access to plentiful high quality forage after a period of rationed feed. Animals that are wintered at low rates of gain and are allowed to compensate in the spring frequently weigh almost the same by mid-summer as those managed through the winter at a high rate of gain. Also known as "pop."

Creep grazing: The allowing of calves to graze ahead of their mothers by keeping the forward paddock wire high enough for the calves to go under but low enough to restrain the cows.

CWT: 100 pounds.

Deferred grazing: The dropping of a paddock from a rotation for use at a later time.

Dirty Fescue: Fescue containing an endophyte which lowers the animal's ability to deal with heat. Fescue without this endophyte is called Fungus-free or Endophyte-free.

Dry matter: Forage after the moisture has been removed.

Flogging: The grazing of a paddock to a very low residual. This is frequently done in the winter to stimulate clover growth the following spring.

Free choice: Non-restricted feeding.

Grazer: An animal that gathers its food by grazing.

Grazier: A human who manages grazing animals.

Green feeding: Direct grazing of corn.

Heavy metal: Large machinery.

K: Potassium.

Lax grazing: The allowing of the animal to have a high degree of selectivity in their grazing. Lax grazing is used when a very high level of animal performance is desired.

Ley pasture: Temporary pasture. Usually of annuals.

Leader-follower Grazing: The use of a high production class of animal followed by a lower production class. For example, lactating dairy cattle followed by replacements. This type of grazing allows both a high level of animal performance and a high level of pasture utilization. Also, called first-last grazing.

Lodged over: Grass that has grown so tall it has fallen over on itself. Most grasses will self-smother when lodged. A major exception is Tall fescue and for this reason it is a prized grass for autumn stockpiling.

Management-intensive grazing or MIG: The thoughtful use of grazing manipulation to produce a desired agronomic and/or animal result. This may include both rotational and continuous stocking depending upon the season.

Mixed grazing: The use of different species grazing either together or in a sequence.

Mob grazing: A mob is a group of animals. This term is used to indicate a high stock density.

N: Nitrogen.

Oklahoma bop: A low stress method of dehorning stocker and feeder cattle whereby a one to two inch stub of horn is allowed to remain. Widely used in the South and Southwest.

P: Phosphorus.

Paddock: A subdivision of a pasture.

Pastureland: Land used primarily for grazing purposes.

Pop: Compensatory gain.

Popping Paddocks: Paddocks of high quality grass and legumes used to maximize compensatory gain in animals before sale or slaughter.

Pugging or bogging: Breaking the sod's surface by the animals' hooves in wet weather. Can be used as a tool for planting new seeds.

Put and take: The adding and subtracting of animals to maintain a desired grass residual and quality. For example, the movement of beef cows from rangeland to keep a rapidly growing tame stocker or dairy pasture from making a seedhead in the spring and thereby losing its quality.

Range: A pasture of native grass plants.

Rational Grazing: André Voisin's term for management-intensive grazing. Rational meant both a thoughtful approach to grazing and a rationing of forage for the animal.

Residue: Forage that remains on the land after a harvest.

Residual: The desired amount of grass to be left in a

paddock after grazing. Generally, the higher the grass residue, the higher the animal's rate of gain and milk production.

Rollback: Light cattle usually sell for a higher price than heavier cattle due to their lower body maintenance. The price spread between light and heavy cattle is called the rollback. See also Value of Gain.

Seasonal grazing: Grazing restricted to one season of the year. For example, the use of high mountain pastures in the summer.

Self feeding: Allowing the animals to eat directly from the silage face by means of a rationing electric wire or sliding headgate.

Set stocking: The same as continuous stocking. Small groups of animals are placed in each paddock and not rotated. Frequently used in the spring with beef and sheep to keep rapidly growing pastures under control.

Split-turn: The grazing of two separate groups of animals during one grazing season rather than one. For example, the selling of one set of winter and spring grazed heavy stocker cattle in the early summer and the replacement of them with lighter cattle for the summer and fall.

Spring flush or lush: The period of very rapid growth of cool-season grasses in the spring.

Standing hay: The deferment of seasonally excess grass for later use. Standing hay is traditionally dead grass. Living hay is the same technique but with green, growing grass.

Stock density: The number of animals on a given unit of land at any one time. This is traditionally a short-term measurement. This is very different from stocking rate, which is a long term measurement of the whole pasture. For example: 200 steers may have a long-term stocking rate of 200 acres, but may for a half a day all be grazed on a single acre. This acre while being grazed would be said to have a stock density of 200 steers to the acre.

Stocker cattle: Animals being grown on pasture between weaning and final finish. Stocker cattle weights are traditionally from 350 to 850 lbs.

Stocking rate: A measurement of the long-term carrying

capacity of a pasture. See stock density.

Stockpiling: The deferment of pasture for use at a later time. Traditionally this is in the autumn. Also known as "autumn saved pasture" or "foggage."

Stripgraze: The use of a frequently moved temporary fence to subdivide a paddock into very small breaks. Most often used to ration grass during winter or droughts.

Sward: The grass portion of the pasture.

Swath grazing: The cutting and swathing of a crop, such as oats, into large double-size windrows. These windrows are then rationed out to animals during the winter by using temporary electric fence. This method of winter feeding is most-often used in cold, dry winter climates.

TDN: Total digestible nutrients.

Value of gain: The net value of gain after the price rollback of light to heavy cattle has been deducted. To find the net value of gain, the total price of the purchased animal is subtracted from the total price of the sold animal. This price is then divided by the number of cwts of gain. Profitability is governed by the value of gain rather than the selling price per pound of the cattle.

Wintergraze: Grazing in the winter season. This can be on autumn saved pasture or on specially planted winter annuals such as cereal rye and annual ryegrass.

Author's Bio

Allan Nation has been the editor of **Stockman Grass Farmer** since 1977. The son of a commercial cattle rancher, Nation has traveled the world studying and photographing grassland farming systems. As a speaker he has been a featured presenter in the United States, Canada, Mexico, Ireland and New Zealand. In 1987, he authored a section on intensive grazing in the USDA Yearbook of Agriculture and has served as a consultant and resource for Audubon Society Television Specials, WTBS, PBS and National Public Radio. He received the 1993 Agricultural Conservation Award from American Farmland Trust for spearheading the drive behind the grass farming revolution in the U.S. He is also the author of **Grass Farmers, Paddock Shift**, and **Quality Pasture**. Allan is married to novelist Carolyn Thornton and they live in Mississippi.

Index

More from Green Park Press

AL'S OBS, 20 Questions & Their Answers by Allan Nation. By popular demand this collection of Al's Obs is presented in question format. Each chapter was selected for its timeless message. 218 pages. **$22.00***

COMEBACK FARMS, Rejuvenating soils, pastures and profits with livestock grazing management by Greg Judy. Takes up where *No Risk Ranching* left off. Expands the grazing concept on leased land with sheep, goats, and pigs in addition to cattle. Covers High Density Grazing, fencing gear and systems, grass-genetic cattle, developing parasite-resistant sheep. 280 pages. **$29.00***

GRASSFED TO FINISH, A production guide to Gourmet Grass-finished Beef by Allan Nation. How to create a forage chain of grasses and legumes to keep things growing year-around. A gourmet product is not only possible year around, but can be produced virtually everywhere in North America. 304 pages. **$33.00***

KICK THE HAY HABIT, A practical guide to year-around grazing by Jim Gerrish. How to eliminate hay - the most costly expense in your operation - no matter where you live in North America. 224 pages. **$27.00*** or Audio version - 6 CDs with charts & figures **$43.00**

KNOWLEDGE RICH RANCHING by Allan Nation. In today's market knowledge separates the rich from the rest. It reveals the secrets of high profit grass farms and ranches, and explains family and business structures for today's and future generations. The first to cover business management principles of grass farming and ranching. Anyone who has profit as their goal will benefit from this book. 336 pages. **$32.00***

LAND, LIVESTOCK & LIFE, A grazier's guide to finance by Allan Nation. Shows how to separate land from a livestock business, make money on leased land by custom grazing, and how to create a quality lifestyle on the farm. 224 pages. **$25.00***

MANAGEMENT-INTENSIVE GRAZING, The Grass-roots of Grass Farming by Jim Gerrish. Details MiG grazing basics: why pastures should be divided into paddocks, how to tap into the power of stock density, extending the grazing season with annual forages and more. Chapter summaries include tips for putting each lesson to work. 320 pages. **$31.00***

More from Green Park Press

MARKETING GRASSFED PRODUCTS PROFITABLY by Carolyn Nation. From farmers' markets to farm stores and beyond, how to market grassfed meats and milk products successfully. Covers pricing, marketing plans, buyers' clubs, tips for working with men and women customers, and how to capitalize on public relations without investing in advertising. 368 pages. **$28.50**

NO RISK RANCHING, Custom Grazing on Leased Land by Greg Judy. Based on first-hand experience, Judy explains how by custom grazing on leased land he was able to pay for his entire farm and home loan within three years. 240 pages. **$28.00***

PADDOCK SHIFT, Revised Edition Drawn from Al's Obs, Changing Views on Grassland Farming by Allan Nation. A collection of Al's Obs as timeless today as when first written. 176 pages. **$20.00***

PA$TURE PROFIT$ WITH STOCKER CATTLE by Allan Nation. Profiles Gordon Hazard, who accumulated and stocked a 3000-acre grass farm solely from retained stocker profits and no bank leverage. Nation backs his economic theories with real life budgets, including one showing investors how to double their money in a year by investing in stocker cattle. 192 pages **$24.95***

QUALTIY PASTURE, How to create it, manage it, and profit from it by Allan Nation. No nonsense tips to boost profits with livestock. How to build pasture from the soil up. How to match pasture quality to livestock class and stocking rates for seasonal dairying, beef production, and multispecies grazing. Examples of real people making real profits. 288 pages. **$32.50***

THE MOVING FEAST, A cultural history of the heritage foods of Southeast Mississippi by Allan Nation. How using the organic techniques from 150 years ago for food crops, trees and livestock can be produced in the South today. 140 pages. **$20.00***

* All books softcover. Prices do not include shipping & handling

**To order call 1-800-748-9808
or visit www.stockmangrassfarmer.com**

Name _____

Address _____

City _____

State/Province_____ Zip/Postal Code _____

Phone _____

Quantity	Title	Price Each	Sub Total
____	**20 Questions** (weight 1 lb)	**$22.00**	_____
____	**Comeback Farms** (weight 1 lb)	**$29.00**	_____
____	**Grassfed to Finish** (weight 1 lb)	**$33.00**	_____
____	**Kick the Hay Habit** (weight 1 lb)	**$27.00**	_____
____	**Kick the Hay Habit Audio - 6 CDs**	**$43.00**	_____
____	**Knowledge Rich Ranching** (wt 1½ lb)	**$32.00**	_____
____	**Land, Livestock & Life** (weight 1 lb)	**$25.00**	_____
____	**Management-intensive Grazing** (wt 1 lb)	**$31.00**	_____
____	**Marketing Grassfed Products Profitably** (1½)	**$28.50**	_____
____	**No Risk Ranching** (weight 1 lb)	**$28.00**	_____
____	**Paddock Shift** (weight 1 lb)	**$20.00**	_____
____	**Pa$ture Profit$ with Stocker Cattle** (1 lb)	**$24.95**	_____
____	**Quality Pasture** (weight 1 lb)	**$32.50**	_____
____	**The Moving Feast** (weight 1 lb)	**$20.00**	_____
____	Free Sample Copy *Stockman Grass Farmer* magazine		_____

Sub Total _____

Mississippi residents add 7% Sales Tax _____

Postage & handling _____

TOTAL _____

Shipping	Amount	Canada	
Under 2 lbs	$5.60	1 lb	$10.00
2-3 lbs	$7.00	1½ lb	$13.50
3-4 lbs	$8.00	2-5 lbs	$20.00
4-5 lbs	$9.60		
5-6 lbs	$11.50		
6-8 lbs	$15.25		
8-10 lbs	$18.50		

Foreign Postage:
Add 40% of order

www.stockmangrassfarmer.com

Please make checks payable to

Stockman Grass Farmer
PO Box 2300
Ridgeland, MS 39158-2300

1-800-748-9808
or 601-853-1861
FAX 601-853-8087